REEDS
VHF
HANDBOOK

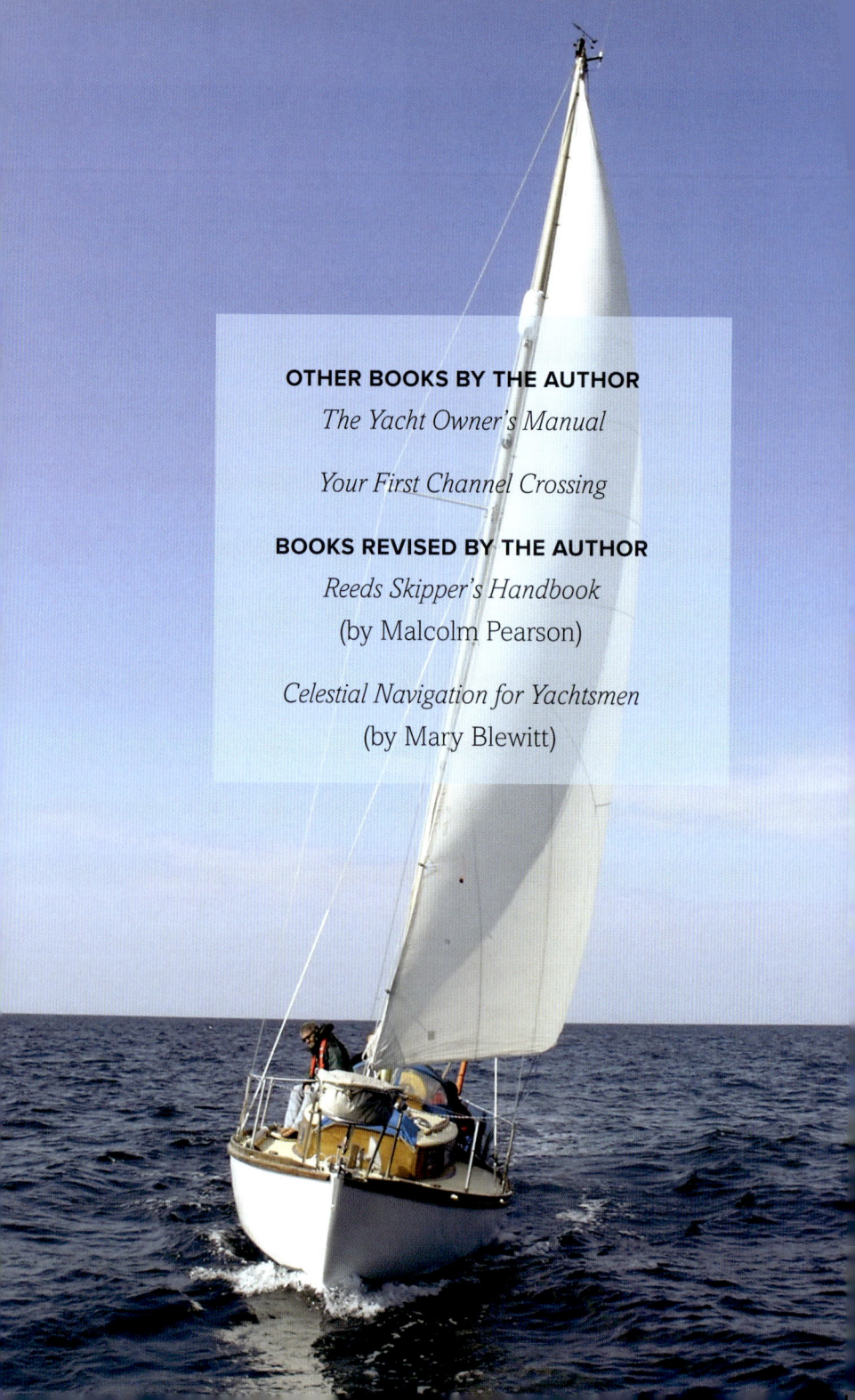

OTHER BOOKS BY THE AUTHOR

The Yacht Owner's Manual

Your First Channel Crossing

BOOKS REVISED BY THE AUTHOR

Reeds Skipper's Handbook
(by Malcolm Pearson)

Celestial Navigation for Yachtsmen
(by Mary Blewitt)

REEDS
VHF
HANDBOOK

Andy Du Port

REEDS
LONDON • OXFORD • NEW YORK • NEW DELHI • SYDNEY

REEDS

Bloomsbury Publishing Plc

50 Bedford Square, London, WC1B 3DP, UK

BLOOMSBURY, REEDS, and the Reeds logo are trademarks of
Bloomsbury Publishing Plc

First published in Great Britain 2021
This edition published 2021

A catalogue record for this book is available from the British Library

Library of Congress Cataloguing-in-Publication data has been applied for.

ISBN: 978-1-4729-8144-8; ePUB: 978-1-4729-8143-1;
ePDF: 978-1-4729-8142-4

2 4 6 8 10 9 7 5 3 1

Typeset in 11/15pt Amasis Light by Richard Carr
Printed and bound in India by Replika Press Pvt. Ltd.

To find out more about our authors and books visit www.bloomsbury.com
and sign up for our newsletters

Contents

Introduction

Thirty or forty years ago it was rare for a VHF (Very High Frequency) radio to be fitted in small craft. Now you would be hard pressed to find a vessel of any size without one. They are routinely carried in sailing yachts, motor cruisers, dinghies, RIBs, kayaks and almost anything else that floats. However, before you can operate one you must hold a Short Range Certificate (SRC).

Whether you are working towards your SRC or you have been using a VHF radio for years, this book is for you. It is laid out so that you can quickly and easily find what you are looking for: from basic radio theory and an introduction to the Global Maritime Distress & Safety System (GMDSS), through to guidance on sending and receiving Routine, Urgency and Distress messages, both by voice and Digital Selective Calling (DSC).

A VHF radio is not only for use in emergencies. You can use it to monitor shipping movements, obtain weather forecasts, arrange your berth in a marina or simply call your chum to decide where to anchor for lunch. Because VHF radios are so widely fitted, the airwaves have become increasingly busy, to the extent that safety can be compromised because essential transmissions are 'lost in the noise'. For this reason there are rules and regulations which we all have to obey. Radios and operators must be licensed, and it is in everyone's interest that standardised procedures are adopted to reduce the length of transmissions and allow urgent messages to be received, understood and acted upon as quickly as possible.

Just as passing your driving test is merely the start of the process of becoming a good driver, obtaining your SRC simply proves that you stayed awake and paid attention during the course. Only by practice and experience will you become competent and confident in operating your radio.

One thing you need to recognise from the start is that a VHF radio is not a telephone. On a phone you are free to chat for as long as you wish and, within reason, say what you want. It is a one-to-one device; a radio is not. As soon as you transmit, not only can everyone within about 20 miles overhear your conversation, you will also be blocking the channel in use until you 'hang up'.

- Throughout the book, specific VHF channels are referred to as 'Ch X' or 'Ch XX' (Ch 6, 8, 16, 67, etc).

- Distress, Urgency, Safety and Routine – with initial capitals – have specific meanings in the context of safety at sea. When in lower case, their everyday meaning applies.

- *Reeds* refers to *Reeds Nautical Almanac*, also published by Adlard Coles.

Explanatory notes are shown with a  symbol in blue-tinted boxes; more general hints and tips are in yellow boxes.

By way of some light relief, at Appendix A there are a few examples of transmissions which are either amusing, include unfortunate wording or simply highlight poor VHF procedure.

Appendices B and C, Check-off list for Distress alerts and Pro forma for MAYDAY calls, are free from copyright and may be copied or reproduced for your own use on board.

The last two pages, for you to complete for MMSI numbers and useful VHF channels, are also free from copyright.

Everything you need to pass your SRC exam and operate your VHF radio with confidence is within these pages. Good luck!

Andy Du Port

Acknowledgements

I am most grateful to Bill Walker for checking the first draft, and to Peter Bailey for reading subsequent drafts and making invaluable suggestions. Also to my daughter Lucy for going through the text with a fine-tooth comb and spotting numerous typos.

Standard Horizon, Icom, Garmin, McMurdo, Digital Yacht, NASA Marine and Navico have kindly given their permission to use their images and diagrams. Other images and photos have kindly been provided by various friends, particularly Philippa Clare from her yacht *Ocean Dancer.*

And the team at Adlard Coles who, as usual, gave sage advice and offered friendly and enthusiastic support and encouragement.

1
Radio Theory

For the purposes of this book, you will be glad to know that there is no need to delve deeply into the theory of radio. Suffice to say, transmitters generate radio waves of various frequencies and wavelengths depending on the purpose of the transmission and the range required. Radio frequencies are sorted into 'bands':

Band designation	Frequency range	Examples of uses
VLF (Very Low Frequency)	3 – 30kHz	Long range comms with submarines
LF (Low Frequency)	30 – 300kHz	Long range comms BBC Radio 4 LW
MF (Medium Frequency)	300kHz – 3MHz	Medium range comms AM radios Navtex
HF (High Frequency)	3 – 30MHz	World-wide comms
VHF (Very High Frequency)	30 – 300MHz	Short range maritime comms
UHF (Ultra High Frequency)	300MHz – 3GHz	GPS, EPIRBs and INMARSAT
SHF (Super High Frequency)	3 – 30GHz	Radar and SARTs

Fig 1 Radio frequency bands

As frequency increases, the wavelength decreases. VLF wavelengths can be 100 kilometres, but SHF wavelengths are only a few centimetres. **VHF has wavelengths of 1 – 10 metres.**

One hertz (Hz) is one cycle per second.
One kilohertz (kHZ) = 1,000Hz, one megahertz (MHz) = 10^6Hz and one gigahertz (GHz) = 10^9Hz.

Range

From Fig 1 you can see that the range of radio waves varies from world-wide to just a few miles. HF waves, for example, may be made to 'bounce' off the ionosphere to allow communications over thousands of miles, while VHF waves are limited to 'line of sight'. This means that they cannot 'see' over the horizon, and range is therefore determined by the heights of the transmitting and receiving aerials: the higher the aerials, the greater the range.

Actually, VHF radio waves do tend to 'bend' around the earth's surface very slightly, so their range is marginally further than the visible horizon.

For all practical purposes, however, the transmitting and receiving aerials need to be in sight of each other. So, if there is high land or other obstructions between the aerials, you will not be able to communicate.

To listen to the Inshore Waters Forecast or the Shipping Forecast on a transistor radio you will need a set capable of receiving Long Wave (LW) and/or FM transmissions. Such radios are readily and cheaply available on the high street, but not all have LW.

The same forecasts are also transmitted as part of the Maritime Safety Information (MSI) broadcasts on VHF.

Fig 2 shows the sort of ranges you should expect in a sailing yacht with the aerial at the masthead (16m is used in these examples) and in a motor yacht with an aerial height of 5m.

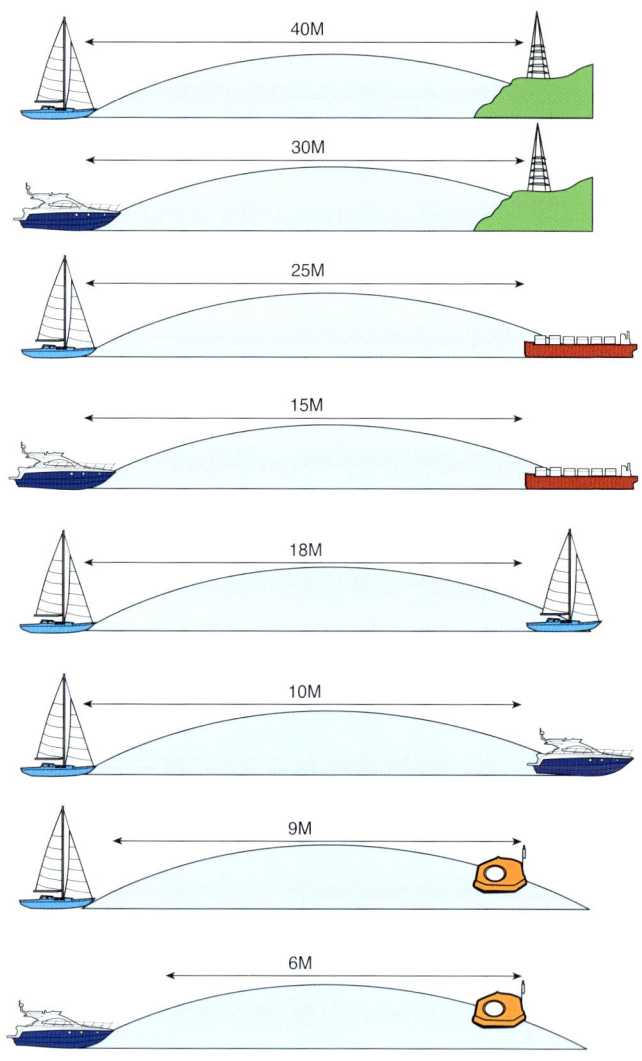

Fig 2 Diagram of VHF ranges

In practice, and in the right atmospheric conditions, you can often achieve rather greater ranges. With a masthead height of 16m, I have frequently passed my passage plan to Solent Coastguard while in Cherbourg or Alderney, a range of about 60 miles.

The other factor affecting range is the power output of the radio. For fixed installations, high power is 25W which is ample to achieve maximum range. Low power (1W) can achieve ranges of 2 – 3 miles and should always be tried before switching to high power. Not only does this prevent your transmissions interfering with other radio traffic at greater distances, it will also reduce battery drain in a handheld set.

Handheld radios

Maximum power output on handheld sets is usually about 6W. Coupled with a low aerial height – typically about 2m if used in the cockpit – the range is unlikely to be more than 3 – 4 miles.

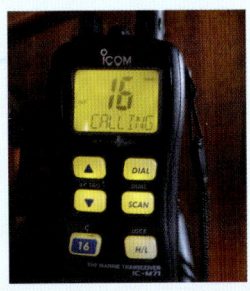

And that is about all you need to know about radio theory. There are plenty of books and online articles which go into much more detail if you are interested. We will now move on to look at GMDSS and where VHF radios fit in to the system.

2
Global Maritime Distress & Safety System

What is GMDSS?

Before ships were routinely fitted with radios the only method of calling for help in an emergency was by visual signals: waving your arms about, hoisting code flags, firing rocket flares or even burning tar barrels on deck. Many of these are still officially recognised Distress signals, although the burning of barrels no longer features – and is definitely not recommended on a yacht!

Even after radios became more commonplace they were not entirely reliable and there was little agreement on operating procedures. For example, in 1906 the now internationally recognised SOS signal was formally adopted, but the Marconi International Marine Communication Company decreed that its operators should continue to use the old Distress signal of CQD. Hence, in 1912 RMS *Titanic* sent a mix of both SOS and CQD Distress signals. Something had to be done, but it was not until 1992 that the Global Maritime Distress & Safety System came into force, under the auspices of the International Maritime Organisation (IMO), with the aim of improving safety of life at sea.

GMDSS comprises many elements working together to form a truly global system using conventional radios, satellite communications, emergency beacons and Digital Selective Calling (DSC). This may sound complicated, but leisure craft are not obliged to carry any of it, and we are really only interested in where VHF radios and DSC fit in to the wider picture.

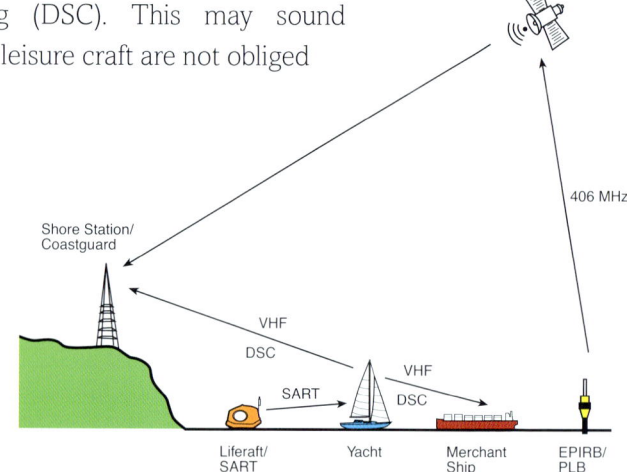

Fig 3 GMDSS

All sea areas are in one of four GMDSS Areas.

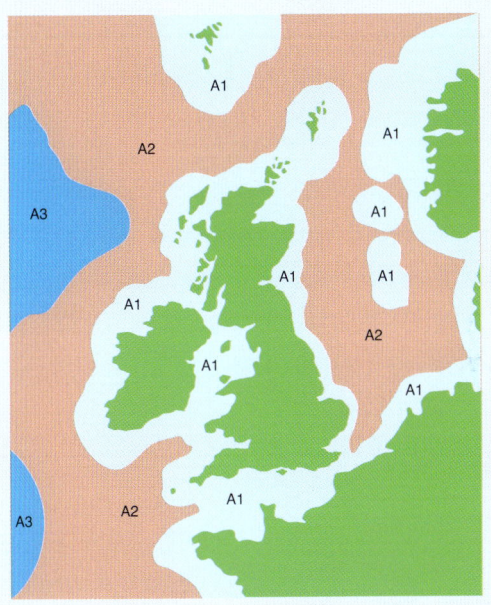

Area A1
Within VHF range of at least one coast station using DSC. In Northern Europe this includes the English Channel, Irish Sea and anywhere else up to about 40 miles offshore.

Area A2
Within MF range of at least one coast station using DSC, out to about 150 miles offshore.

Area A3
Within coverage of INMARSAT satellites (70°N to 70°S).

Area A4
All other areas, including the polar regions, using HF.

Fig 4 GMDSS coverage

Most of us operate in Area A1, but you can see from Fig 4 that much of the North Sea and some of the Western Approaches to the English Channel fall into Area A2. This is not a problem in practice as in those areas you will always be within VHF range of at least some of the hundreds of ships which ply those waters every day of the year.

Satellite systems associated with GMDSS

Cospas/Sarsat is an international network of satellites which relay transmissions from distress beacons such as EPIRBs and PLBs (more about these later) to Search and Rescue (SAR) authorities back on Earth. The system can receive signals from distress beacons which transmit on 406MHz. Older beacons which transmit on 121.5MHz or 243MHz rely on being received by aircraft and associated rescue assets. They are not compatible with Cospas/Sarsat unless they also transmit on 406MHz.

> If you really want to know, COSPAS is a Russian acronym for **CO**smicheskaya **S**isteyama **P**oiska **A**variynich **S**udov; SARSAT, rather more comprehensibly, stands for **S**earch **a**nd **R**escue **S**atellite-**A**ided **T**racking.

INMARSAT is a British satellite communications system which covers all areas except the polar regions and can be used to transmit distress messages. It is thus part of GMDSS, but for our purposes it is of passing interest.

Other satellite systems. You may also come across Galileo, LEOSAR, GEOSAR, MEOSAR and others. They are all part of the architecture of the Cospas/Sarsat satellite system and need concern us no further.

NAVTEX

Navtex is an integral part of GMDSS and provides navigational and weather information, in English, out to at least 200 miles offshore; some stations have a range of up to 400 miles. Broadcasts are best received using a dedicated receiver which has its own aerial. Information may either be displayed on a screen or as a paper printout.

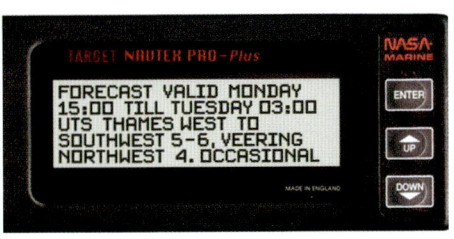

The great advantage of having a Navtex receiver on board is that you no longer have to worry about missing a weather forecast (or falling asleep at the critical moment). That said, if your sailing is mainly coastal with just the occasional passage further offshore, you should be able to get perfectly adequate Marine Safety Information (MSI) broadcasts from coast radio stations, either on your VHF radio or, as a back-up when further offshore, with a portable radio capable of receiving Long Wave (LW) transmissions. There is also a plethora of online weather forecasts to choose from.

Details of Navtex stations, messages and message categories may be found online or in *Reeds*. Each station has a single-letter identification followed by the message category and a two-digit serial number of the message.

Navtex receivers suitable for yachts are readily available and cost between about £150 and £300.

Navtex message identification 'EA12'
E = Niton; **A** = Navigation warning; **12** = Serial number. A serial number of '00' denotes urgent traffic such as a gale warning.

EPIRBs, PLBs and SARTs

The last three elements of GMDSS which are relevant to us are locator beacons.

EPIRB (**E**mergency **P**osition **I**ndicating **R**adio **B**eacon) transmits a Distress alert on 406Mhz including its identity, which is linked to your vessel, and your position which is either from an integrated GPS receiver or calculated by the Cospas/Sarsat satellites. The latter takes rather longer to get a fix.

An EPIRB may be mounted on a bracket which automatically releases and activates the beacon as the vessel sinks (Category 1), or it must be manually removed by hand before it can be activated (Category 2).

If you activate your EPIRB accidentally, you must turn it off immediately and inform the Coastguard as soon as possible, giving as much detail as you can.

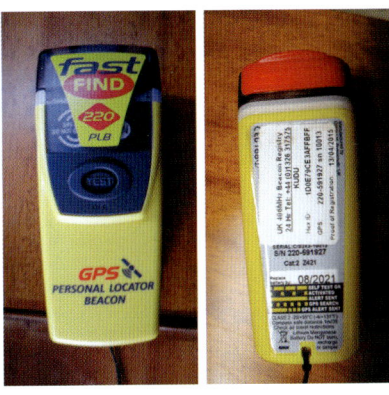

PLB (**P**ersonal **L**ocator **B**eacon) is basically a personal EPIRB. It is registered to a person rather than a vessel, has a shorter battery life (about 24 hours), has to be activated manually and is significantly smaller and cheaper.

For yachts which don't make long offshore passages, a PLB is worth considering. It can be attached to your lifejacket and activated if you should fall overboard; it can be in your emergency grab bag; or it can be lashed to the liferaft.

All EPIRBs and PLBs must be registered with the UK Beacon Registry:

https://www.gov.uk/maritime-safety-weather-and-navigation/register-406-mhz-beacons

SART (**S**earch **a**nd **R**escue **T**ransponder) is primarily intended for use in a lifeboat or liferaft to pinpoint its position. It should be mounted as high as possible to maximise range.

When activated, the transmitted signal shows on a radar screen as 12 dots radiating from the

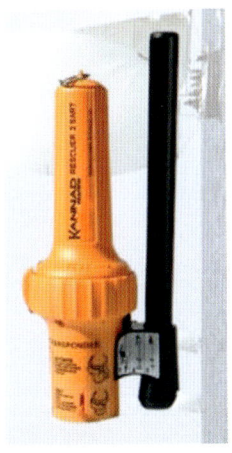

A SART may be GPS/AIS enabled. It will then display its position on an AIS receiver.

SART's position. As you get closer to the transponder, the dots expand into arcs and eventually, when very close, into concentric circles. Detection range by surface ships is about 8 miles, while the range from aircraft will be considerably further.

 SARTs are quite expensive and not, as far as I know, carried in many yachts which are sailed mainly in coastal waters.

DSC

DSC is undoubtedly the most significant enhancement to VHF radio in recent years but many small craft skippers, perhaps thinking it should only be used in an emergency, seem to be reluctant to recognise its full potential. Properly used, DSC can greatly reduce the current over-reliance on VHF Ch 16 for making Routine calls. What follows is a brief description of how DSC works. We will look at the practical aspects of using DSC in Chapter 7.

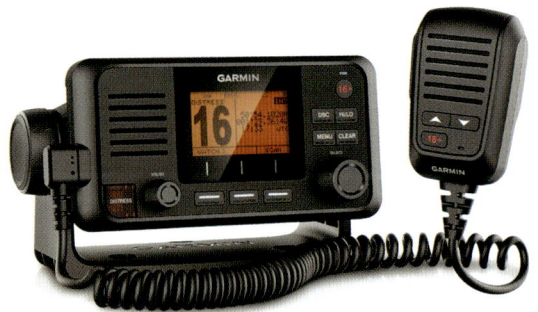

DSC does not increase the range of your radio; like any other VHF transmission, it is limited to line of sight.

A VHF DSC radio takes on many aspects of a telephone: you need to 'dial' the nine-digit number you want to call (which can be stored in a directory in your radio). The radio you are calling then 'rings' until it is answered and you start your conversation on a pre-selected channel.

The DSC function in a handheld radio must not be used outside UK territorial waters

MMSI

The nine-digit 'telephone number', known as the Maritime Mobile Service Identity (MMSI), is unique to your vessel, not the radio. Unless you already hold a Ship Radio Licence (see Chapter 4), when you install a radio (new or second hand), you will need to apply to Ofcom for an MMSI. If you are obtaining a Ship Radio Licence for the first time an MMSI will be included automatically. If the boat is sold the MMSI stays with her, but if you sell or change the radio the old MMSI must be erased. This can only be done by an authorised radio engineer.

Entering the MMSI for the first time requires particular care as you only get a limited number of attempts before the radio is inhibited and will have to be reprogrammed by a radio engineer.

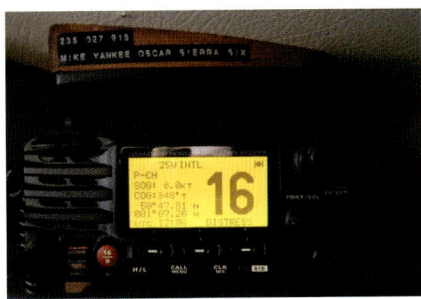

The first three digits of an MMSI indicate the national code: 232, 233, 234 and 235 for the UK. The remaining six digits are unique to your boat. A typical MMSI might be 235 436 067 and is like an electronic callsign.

Coast stations' MMSIs begin with 00 followed by a national code (as above) and a further four digits for the individual station. For example, Dover Coastguard's MMSI is 00 232 0010. MMSIs for coastguard and other shore stations are listed in *Reeds* and the Admiralty List of Radio Signals (ALRS).

 Handheld VHF DSC radios have slightly different rules. They are allocated a separate MMSI which stays with the radio, not the vessel. If you have such a set, you must apply for a Ship Portable Radio Licence.

Group MMSI

Clubs and other organisations may apply to Ofcom for a Group MMSI which is shared by all their members. It is similar to an Individual MMSI, but it does not identify any particular vessel. It consists of an initial zero followed by the three-digit country identifier and five more digits. An example is 0 232 12345.

A Group MMSI is entered into your MMSI directory in the same way as an Individual MMSI; it may be added or deleted as you wish.

DSC transmissions

When you make a DSC call, the radio transmits a very short burst of code on Ch 70 which contains its identity (MMSI) and your position. The radio should therefore have a GPS input, either via its own antenna or from the boat's main GPS set. If that is not possible you

can enter the position manually, but you will need to do this frequently and regularly if it is to be of any use during an emergency. When the DSC signal is received, it triggers an alarm, or 'ring', in the radio you are calling.

 VHF Ch 70 is dedicated to DSC and must never be used for voice communications. Also, note that the use of DSC is restricted in most countries to coastal waters; DSC transmissions should not be made when navigating inland waterways.

AIS

AIS (**A**utomatic **I**dentification **S**ystem) is not yet a part of GMDSS but can make a significant contribution to safety at sea. An increasing number of pleasure craft are equipped with AIS, either just a receiver or a transponder which also transmits. Some VHF radios have integrated GPS and AIS receiver, and are well worth considering if either space or budget is tight.

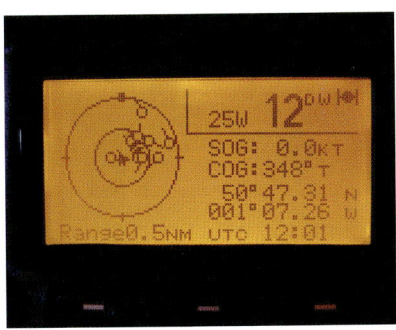

VHF with AIS display

A few manufacturers offer VHF sets which incorporate an AIS transponder. However, they come at a price.

Mainly used for collision avoidance, an AIS transponder transmits, as a minimum, the vessel's identity, position, course and speed. With appropriate connections, these outputs may then be displayed on a chartplotter, laptop or tablet. AIS may, therefore, be a valuable location aid in an emergency. However, Class B AIS (see below) has a power output of only 2W, so the range is less than you would expect from your VHF radio.

Class A AIS is fitted to ships and other commercial vessels of greater than 300 tons, and transmits additional data such as rate of turn, destination and ETA.

Class B, with limited outputs, is mainly for use in small vessels.

VHF with AIS display

 AIS is not a radar, whatever some manufacturers may say! Remember that many small vessels are not equipped with AIS transponders and so they will not show up on your display. AIS is an extremely useful aid to collision avoidance but is no substitute for radar.

The MCA has some wise words on the use of AIS. They are worth repeating here:

Collision avoidance must be carried out in strict compliance with the Colregs. There is no provision in the Colregs for use of AIS information. Therefore, decisions should be taken primarily on systematic visual and/or radar observations. The availability of AIS data similar to that produced by systematic radar-tracking (eg ARPA [or MARPA]) should not be given priority over the latter. AIS target data will be based on ... course and speed over the ground whilst for Colregs compliance such data must be based on ... course and speed through the water.

Good advice.

ATIS

Automatic **T**ransmitter **I**dentification **S**ystem is an extension of AIS and is *required* for vessels making VHF transmissions while on European inland waterways. Having checked that your radio is compatible, you need to apply to Ofcom for an ATIS number. This is your MMSI preceded by '9' (eg 9235 123 123). The identity of your vessel is transmitted digitally every time you release the transmit button.

 Not all VHF sets are capable of ATIS transmissions; you may need to contact your manufacturer for advice.

3
VHF Radios

Having briefly covered radio theory, GMDSS and AIS, we will now look at what VHF radios are available. Although specific makes are shown in the illustrations, I am not recommending any particular manufacturer. That is entirely your choice and will depend on your own preferences and recommendations from others. In practice, all VHF radios share many of the same features and nomenclature. Even the layout of the controls can be very similar.

A relatively new addition to the VHF market is the 'smart' radio which boasts an AIS transceiver, intuitive touchscreen operation, connectivity with smart phones, wi-fi, combined AIS and VHF display, GPS and more. It's probably more suited to large pleasure craft and comes at a price. You will not get much change from £2,000 for a 'basic' version.

All new VHF fixed radios include DSC but some older sets, and many handheld radios, do not. If your boating is mainly in local waters and your only requirement is to make an occasional call to your harbour master or marina or, of course, in the event of an emergency, you may opt for a relatively basic set or even just a handheld radio.

If you sail further afield and are likely to be in busy shipping areas such as the North Sea or English Channel, then the inclusion of AIS could be invaluable. In this case, a set which also has its own GPS receiver is a neat solution and makes for simple installation.

Not many VHF radios yet include an AIS transponder (receive *and* transmit), and they are quite expensive. However, you may feel that this offsets the need for a stand-alone AIS transceiver and all the associated connectivity.

Prices vary from less than £150 to well over £1,000. A good DSC radio with built-in AIS (receive only) might set you back about £300. As ever, you tend to get what you pay for. A 'bargain' radio may not be as robust or capable as more expensive alternatives.

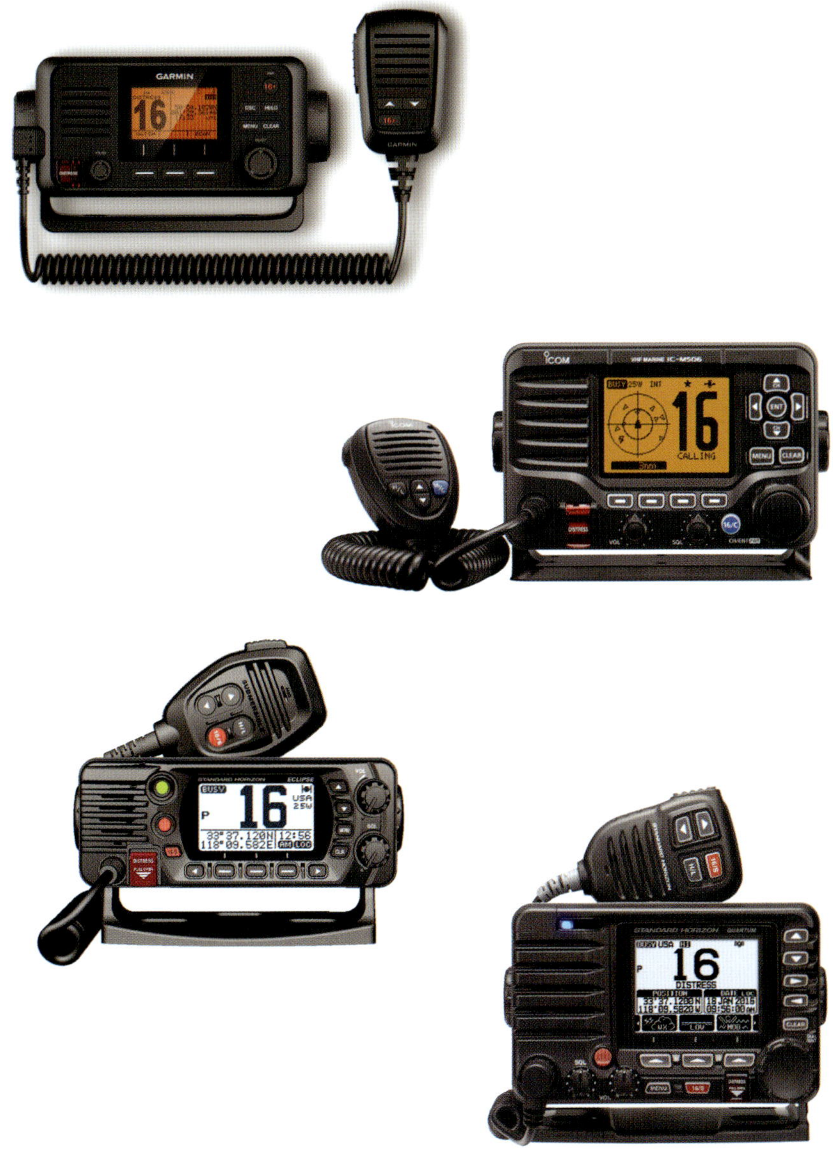

Radios with various features.

Fig 5 Diagram of a VHF radio controls

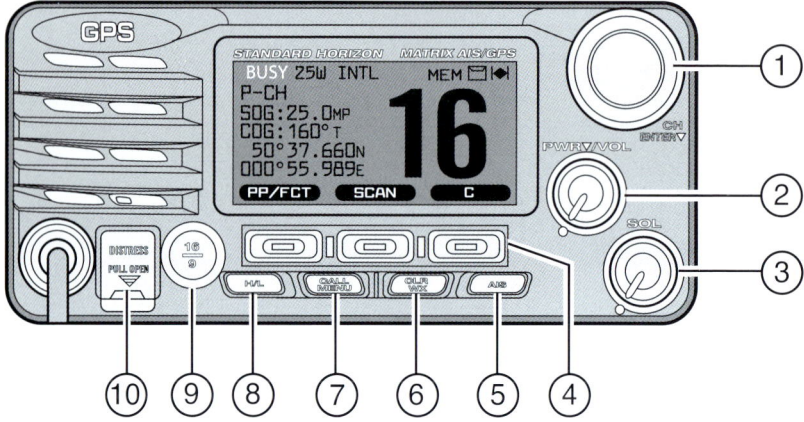

VHF radios typically have the following controls:

1 Channel/Enter Either a rotary switch or a keypad to select the VHF channel. **Enter** enables MMSIs, manual positions and any other text-based functions to be inserted. Also used (in this example) to confirm or select a channel or other menu item.

2 Power/Volume This may be a rotary switch combined with the volume control, or it may be a 'press and hold' button. Used to turn the radio on or off and to control the sound level of the internal speaker and any remote speaker.

3 Squelch The squelch control affects the radio's sensitivity and is used to filter out interference ('hiss') which may affect the signal you want to hear. It should be used with caution – just enough to cut the noise but no more.

4 Soft keys Many sets have three or more 'soft keys' which appear on the screen for assorted menus or functions.

Scan On the radio illustrated, the scan function is selected with a soft key; on others there may be a dedicated 'Scan' button. Three scan options are usually available: **dual-watch** (for monitoring Ch 16 and one other), **memory scan** (as many channels as you select) and **priority scan** which is similar to memory scan but the radio checks Ch 16 between each selected channel. The radio will lock on to a received channel for as long as it is transmitting (or for a specified time) before resuming the selected scan function.

5 AIS Selects the AIS screen.

6 CLR Clears a previous entry.

7 Call/Menu For making non-Distress DSC transmissions and for displaying menus on the screen.

8 Hi/Lo Used to switch between high power (25W) or low power (1W).

9 Ch 16 VHF Ch 16 is the Distress and calling channel, and this button allows you to access it instantly when necessary. When pressed it should also automatically select high power.

10 Distress This will be a red button under a cover to prevent accidental use. It often requires two hands to operate. The use of this feature is covered in detail later.

PTT, or press to transmit, button is on the microphone or handset (not shown on the diagram on page 36). It does exactly what it says: you press it to transmit and release it to receive. It is important to make sure the button does not remain inadvertently pressed as your radio will then send a continuous signal and you will not be able to receive messages.

It really is essential to break the habit of a lifetime and read the manual. Most radios have so many features and settings that it is prudent to keep the instruction book close to hand.

The screen displays a plethora of information which you can usually tailor to meet your needs. It will always show the chosen channel group, the channel in use and whether high or low power is selected. It will also indicate if the radio is busy transmitting (TX) or receiving (RX). If GPS data is available, your position and the current time will also be shown.

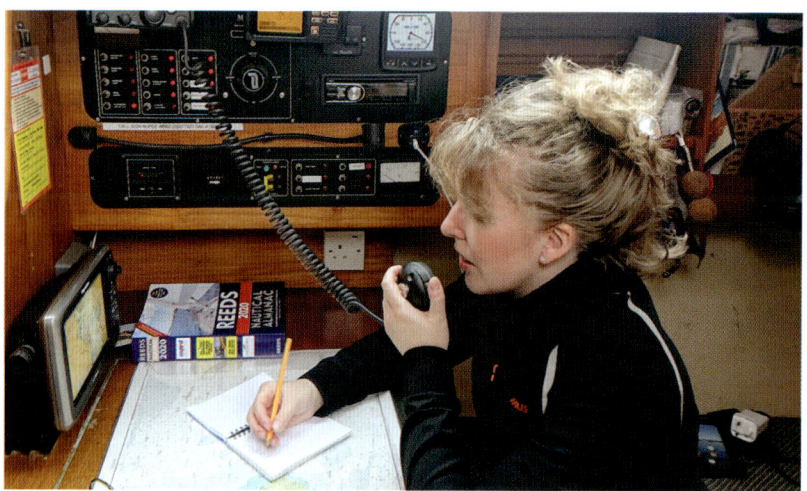

Radio installed at the chart table

Radio installed in the wheelhouse

Siting the radio

Your radio should be sited so that you can easily read the display and operate the controls comfortably. It should not be too close to the steering compass (usually about 1 metre) nor subject to excessive heat or vibration. There are basically two options: at the chart table/ nav station or in the cockpit/steering position.

I prefer to have my radio at the chart table so it can be used in conjunction with other navigational aids, instruments and the chart. The downside is that you have to go below to use it. This can be overcome either by having a handheld set in the cockpit or a remote

handset with which you can select channels and access other functions. Whatever you decide, an extension speaker in the cockpit is, in my view, essential. It goes without saying that the radio must be protected from rain and spray unless it is waterproof (most are).

 For maximum range, the VHF aerial should be installed as high as possible, preferably at the top of the mast, clear of obstructions. AIS transponders only radiate 1 watt, so range is much reduced if the aerial is low down.

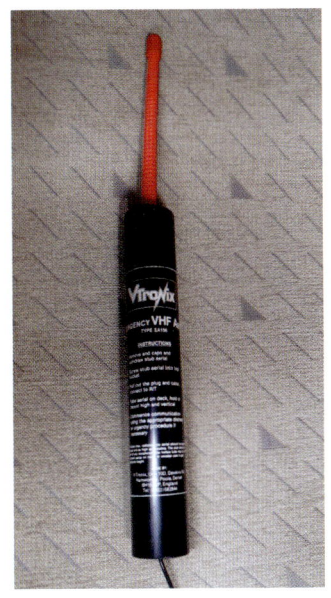

In the event of a dismasting or loss of the VHF aerial, it is a good idea to carry an emergency aerial which can be plugged in to the fixed radio and hoisted as high as possible. This will increase the range significantly as you call for help.

Handheld radios

A very basic handheld radio will cost you up to about £100, but if you opt for a top-end model with GPS, DSC and AIS, you could be looking at £300 or more. A reliable, sturdy, waterproof handheld (under about £150) is fine for use in the cockpit for monitoring traffic, calling marinas, etc.

A handheld set can be taken with you in the dinghy or, in extremis, the liferaft. It is not dependent on the boat's power supply, so is a good back-up to the fixed radio should you suffer a power failure. Be sure to keep it fully charged.

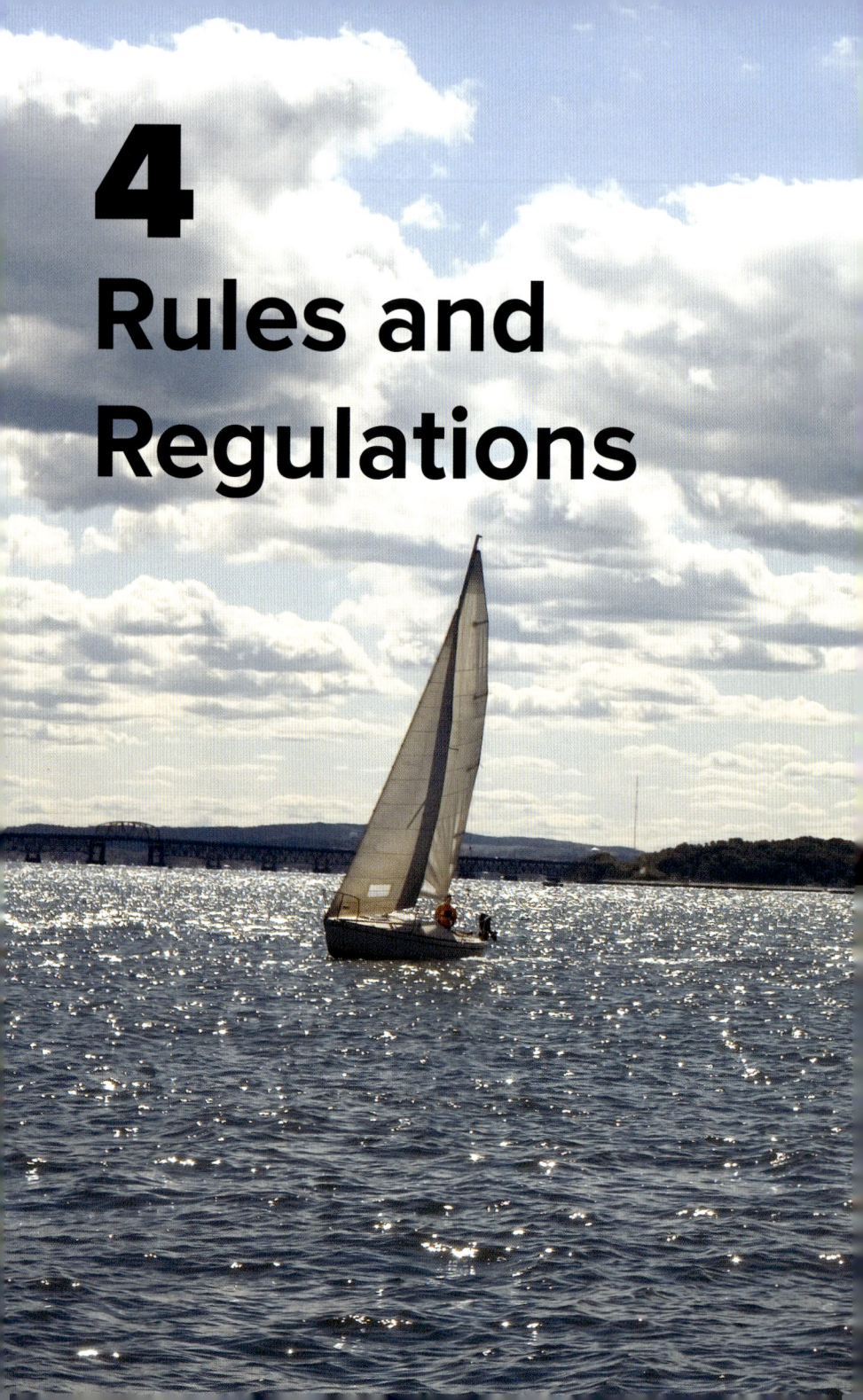

4
Rules and Regulations

So far we have looked at VHF radio in fairly general terms, and where it fits in to GMDSS. We now need to get to grips with the licencing requirements. Below is just a summary of what is required; much more detail can be found on the Ofcom website. Search for *Of168a – Guidance notes for licensing* where you will find all you need to know in an easily digested format.

Ship Radio Licence

This allows you to install and use specified radio equipment in a particular vessel. It is issued by Ofcom, costs nothing and lasts for the life of the vessel, but it is *not* the authority for you to operate the equipment. That's covered by the SRC (see below).

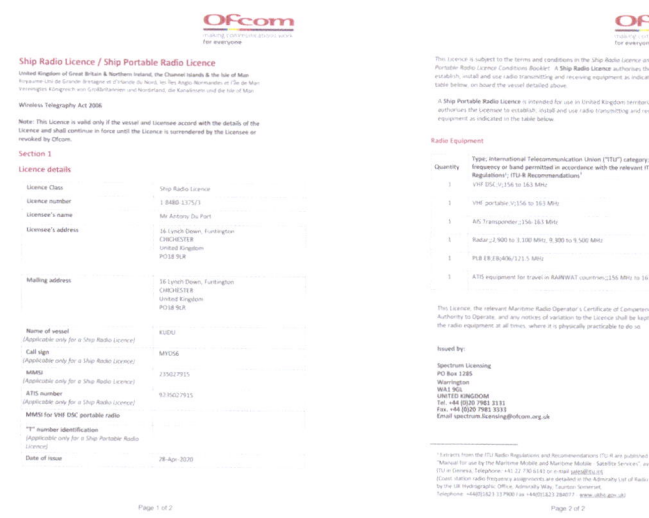

VHF Ship Radio Licence

Although you need a licence for a VHF radio, any other 'maritime radio equipment' on board must also be included:

MF, HF and VHF radios
Satellite communications equipment
AIS transponders
Radar
Handheld VHF or UHF radios
EPIRBs
SARTs

The Ship Radio Licence is issued to you by name, even though it relates to the radio equipment in a particular vessel. This means that you must inform Ofcom if you cease to be the owner or if there are any changes to the equipment carried on board. There are no charges for this.

 It is a requirement that your Ship Radio Licence is carried on board at all times.

Ship Portable Radio Licence

You will need this additional licence if you have a handheld DSC VHF radio, a PLB or an EPIRB which may be used on other vessels. Unlike the Ship Radio Licence, the equipment listed in the Portable Radio Licence is not tied to a particular vessel and is not allocated a callsign. Instead, you will be issued with a 'T-number' which identifies the licensee (ie you) rather than your boat. An MMSI may also be allocated for a DSC handheld radio, but its format makes it clear that it is a portable set.

A Ship Portable Radio Licence is only valid in UK territorial waters.

Short Range Certificate

Your SRC is also your **Authority to Operate** and you must hold this before you use, or supervise the use of, any VHF radio. Anyone may operate the radio, but if they don't hold an SRC they must be directly supervised by someone who does.

The SRC is administered by the RYA (not Ofcom). To obtain your certificate you must take a short exam which involves a practical test (using a dedicated training radio) and a short written paper. By far the easiest way of doing this is to attend a recognised course which will include the exam at the end. The total learning time is 10 hours, but some of the work (up to 3 hours) can be done using a self-learning programme before attending the course in person.

Contact your local RYA training centre for details of when courses are run, or go to the RYA website and search for: *Courses and Training > Find a course.* At the time of writing (2020) the fee for the course, exam and certificate is £60.

There is no minimum age for attending the course, but candidates must be 16 or over to take the exam.

 Important note: all these licences are compulsory; you could face a substantial fine if you don't comply.

Radio log

Although leisure craft are exempt from the requirement for larger vessels to maintain a record of all Distress, Urgency and Safety messages sent or received, it is a good idea to make a note of such events in the boat's log.

5
VHF Channels

As we saw in Chapter 1, the frequency range for VHF is 30MHz to 300MHz. The frequencies between 156MHz and 162MHz are used for communications, and these are allocated to numbered channels: 01 – 28 and 60 – 88.

Most of these channels are duplex; the remainder are simplex.

Duplex channels use two frequencies, one to transmit and one to receive. This means that transmissions are possible simultaneously in both directions – rather like a telephone. However, duplex radios require two aerials and are rarely fitted in pleasure craft. Instead, most of us have 'semi-duplex' radios which switch between the two frequencies. When you call a marina on Ch 80, for example, you will only hear the marina's transmissions, not those of other boats which are also trying to arrange their berths.

Simplex channels use the same frequency for both transmission and reception. The inter-ship channels are simplex: you press the PTT button to send your message and release it to receive. You can thus hear both sides of conversations between other vessels. This is particularly useful when monitoring harbour control and VTS (Vessel Traffic Service) channels.

 There are three VHF channel groups: International, USA and Canada. Unless you are in North America, you should select the International group.

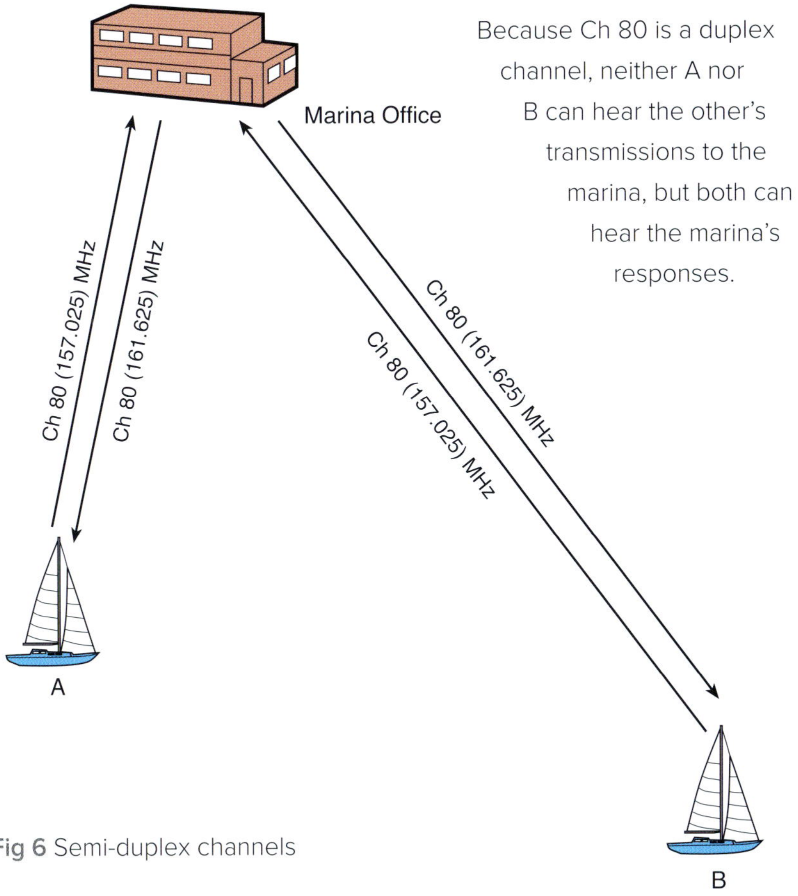

Marina Office

Because Ch 80 is a duplex channel, neither A nor B can hear the other's transmissions to the marina, but both can hear the marina's responses.

Ch 80 (157.025) MHz

Ch 80 (161.625) MHz

Ch 80 (161.625) MHz

Ch 80 (157.025) MHz

A

B

Fig 6 Semi-duplex channels

Ch 6 (156.300 MHz)

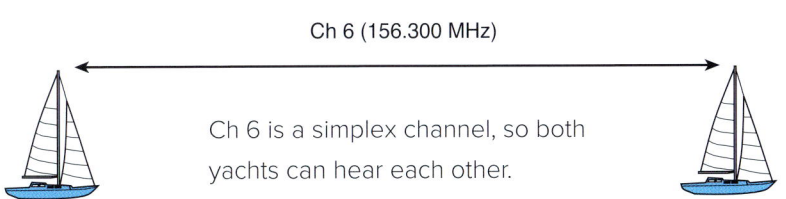

Ch 6 is a simplex channel, so both yachts can hear each other.

Fig 7 Simplex channels

Channel allocation

Channels are allocated according to purpose. We are chiefly concerned with those channels used for emergencies, safety (including weather forecasts), inter-ship communications, port operations and communicating with coast stations.

Ch 16 is the primary *Distress, Urgency and Safety* channel. It is also often used for making an initial call before changing to a working channel for routine traffic. It is a very busy channel and should never be used to pass non-urgent messages. Ch 16 should be monitored by all vessels when at sea.

Ch 13 is for *bridge-to-bridge* communications for navigational safety. Commercial ships should monitor Ch 13, but it is sometimes necessary to make an initial call on Ch 16.

Ch 67 is the *UK Small Craft Safety* channel and is often used when communicating with the coastguard. Solent Coastguard prefers yachts initially to call on Ch 67 instead of Ch 16, especially during busy periods.

 Other Coastguard stations may use different working channels (check in *Reeds*), but make your initial call on Ch 16.

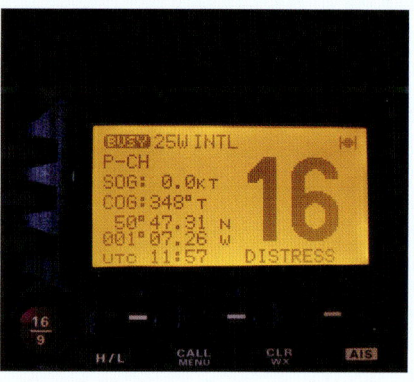

Ch 10, 62, 63 and 64 are used for *MSI broadcasts* around the UK. See *Reeds* to find out which channel covers your area.

Ch 6, 8, 72 and 77 are the main channels for routine *inter-ship* traffic. You should change to one of these channels after establishing contact on Ch 16. Other channels are available; see *Reeds* for more information.

The MCA's advice on using VHF for collision avoidance is as follows:

The use of VHF to discuss actions to take between ships is fraught with danger and is discouraged. The MCA's view is that identification of a target by AIS does not completely alleviate the danger. Decisions on collision avoidance should be made strictly according to the Colregs.

Ch 80 is used in the UK for communicating with many marinas.

Ch 70 *is allocated exclusively to DSC and must not be used for voice communications.*

There are three additional channels:

Ch 0 is used by the coastguard and emergency services. It is not available for general use.

Ch M1 was called Ch 37, but no longer. It is used to contact some UK marinas, yacht clubs, etc. It must only be used in UK territorial waters. It is usually referred to simply as 'Channel M'. Some radios display it as Ch P1.

>  Always listen for a few seconds on your chosen channel to make sure it is not in use before transmitting.

Ch M2 is used mainly by yacht clubs for race control, safety boats, etc. Like Ch M1, it must only be used in UK territorial waters. Some radios display it as Ch P2.

You will find a full list of channels and their purposes in Chapter 14. They will probably also be listed in your radio's user manual.

Which VHF channel should I use?

In the table opposite, you can quickly find the appropriate channel for your purpose.

Purpose	Channel(s)	Notes
Distress/ Urgency	16	Use DSC for initial alerts if possible.
Initial call	16	Keep it short and switch to a working channel as soon as you can. Use an alternative calling channel if available (see below).
Small craft safety	67	Some coastguard stations may prefer to be called directly on Ch 67 or another working channel – check in *Reeds*.
Navigational safety ('bridge-to-bridge')	13	If no response, try calling on Ch 16 before changing to Ch 13 or another working channel.
Working channels ('ship-to-ship')	6, 8, 72, 77	Even on a working channel, keep transmissions as short as possible.
Marinas	80	Some marinas use Ch M or another designated channel – check in *Reeds* before calling. Do *not* use Ch 16.
Vessel Traffic Services (VTS)	12, 14	Check in *Reeds*.
MSI broadcasts	10, 62, 63, 64	Initial announcement is made on Ch 16. The channel number you select depends on the preferred aerial, normally the closest. Check in *Reeds*.
Yacht races/ safety	M2	Other channels may be designated.

6
Radio Procedures

U nless you are very unlucky, you will mostly use your radio for non-emergency 'Routine' traffic. So that is where we will start.

 All the following advice applies equally to Distress, Urgency and Safety traffic.

Golden guidance

Is your transmission necessary? Do you *really* need to call your chum to discuss your plans for lunch? If the usual ship-to-ship channels are busy, you could drown out more urgent traffic. See below for guidance on checking your radio.

Listen Before transmitting listen to Ch 16 for a few seconds before making your initial call to be absolutely certain that you are not about to interrupt Safety traffic.

Keep it brief Don't hog the airwaves. Decide what you want to say beforehand and finish your conversation as quickly as possible. Imagine you are paying by the second!

Be clear Hold the microphone 2 – 5cm from your mouth and slightly to one side. Speak clearly and relatively slowly to make sure you are understood first time. If wind noise is a problem, either go below or turn your back to the wind.

Use low power If the station you are calling is within a couple of miles, try low power. On high power, everyone within about 20 miles can hear you.

Radio checks *Most radio checks are quite unnecessary and should be avoided.* You hardly ever hear merchant ships or fishing vessels conducting a radio check, so why do yachts?

- Modern VHF radios are extremely reliable.
- If your radio is receiving, the aerial run is probably sound.
- A satisfactory radio check is no guarantee that it will work next time. If a fault develops, it will inevitably strike without warning.
- Radio checks can interrupt or interfere with more urgent traffic.
- In a busy yachting area, continuous radio checks can be very irritating.

That said, your radio might be a life-saver, and you should of course have confidence in it. If you have just installed the set, done some work on the aerial run or have good reason to suspect a fault, then by all means make a check. Otherwise, trust the technology!

See Chapter 8 for guidance on conducting a radio check.

Rules

Inevitably, there are some rules which you *must* obey:

- You must identify yourself by using your vessel's name, callsign or MMSI.

- Unofficial 'callsigns' (eg name of the skipper) are not permitted.

- Anyone using the radio must either hold an Authority to Operate (ie an SRC) or be closely supervised by someone who does.

- You must not make false Distress alerts or Urgency announcements. If one is sent in error – perhaps by an errant child – you must cancel it as soon as possible (see Chapter 9).

- If you are directly involved in Distress, Urgency or Safety traffic, you must keep your radio on the relevant working channel until stood down or the incident is resolved.

- You must not 'broadcast blind'. All transmissions must be addressed to a particular station or stations, even if that is to 'All Stations'.

- Do not use obscene, profane or indecent language.

- Do not use VHF to call someone ashore other than an authorised station such as a marina. You must not allow your crew to go ashore with a handheld VHF radio for the purposes of communicating with you in your boat.

As an aside, some years ago I called the coastguard to send a passage report from the yacht *Mallard*, and was asked to spell her name. I did so, and ended – to be helpful – with, 'Like a duck'. This was misheard by the (female) coastguard and there was a rather embarrassing interlude until she understood what I was saying!

Use only those frequencies/channels authorised in your Ship Radio Licence.

It is illegal to divulge the contents of radio messages which are not addressed to you. Common sense is required here. The purpose of this rule is to prevent you from using overheard information for personal gain. Distress communications and transmissions for general use (MSI broadcasts, etc.) are obviously exempt, but you must not repeat anything you overhear that you would want to keep secret yourself.

Prowords

These are words with specific meanings during radio communications. They are designed to be easily understood and unambiguous. They also help to keep transmissions short.

This is Indicates who is calling. You should say:
> *Seabird **this is** Voyager, over.*

Do *not* say:
> *Hello Seabird, Voyager here, over.*

Received *I have received and understood your message.*

You will often hear 'Roger' used instead of 'Received'. It is now out of date and does not feature in current guidance. However, old habits die hard, and some of us occasionally still say 'Roger'. There is no real harm in doing so if, as always, your meaning is obvious – and you don't use it during your SRC exam.

Over *I have finished transmitting; I am now waiting for your reply.*

Out *End of working* [ie your conversation]; *I do not expect a reply.*

It is illogical, but heard only too often, to say '*Over and out*'. You should *never* say this as it would mean, '*I am inviting you to reply, but I'm not listening*'!

Say again *'Repeat your message'*. Usually used when a message is distorted or not understood. It may be used to ask for a repetition of just part of a message:
> *Say again all after …*
> or
> *Say again all before …*

I say again Used after a request to repeat your message (or part of it) or to emphasise a word or phrase:
> *I say again …*
> *I say again all before/after …*
> *Your berth is Alpha 6. I say again, Alpha 6. Over.*

I spell Used in reply to a request to spell a particular word. Use the phonetic alphabet (see Chapter 12) and say the word you are spelling at the end:
> *I spell: Charlie Oscar Whisky Echo Sierra, Cowes. Over.*

Station calling Used when you are unsure of the identity of the calling station:
> *Station calling Voyager, say again your name. Over.*

Correction As in:
> *My ETA is 1130. Correction, 1230. Over.*

In the next chapters, we shall see how to send typical Routine, Urgency and Distress messages, both by voice and by DSC.

7
Using DSC

If you and the station you want to call are both suitably equipped, and you know the MMSI of that station, you can reduce traffic on Ch 16 by making initial contact by DSC. Not many yachts seem to do this for Routine traffic but it is a simple process, and they should.

In the following chapters, you will find further advice on using DSC for Distress, Urgency and Safety calls.

Making a Routine DSC call

Radios vary, so check the user manual for your set.

- Select the DSC menu.

- Select 'Individual' (other options will probably be: Distress, All ships and Group).

- Input the MMSI of the vessel you want to call. [1]

- Select 'Routine' (rather than 'Urgency' or 'Safety').

- Select the working channel you want to use. [2] [3]

- Select 'Transmit' and wait for an acknowledgement 'ring'.

- Press 'Accept' (or whatever your radio requires).

- Pick up your handset and continue your conversation by voice.

This may seem to be quite an involved process but, with practice, it will become second nature to you.

If you are fitted with AIS, you will be able to see the MMSIs of all vessels in range. This is particularly useful if you want to contact a ship to make sure she has seen you (but heed the warning in Chapter 6). You can use DSC to call the coastguard but, particularly if you are calling directly on a working channel, DSC has no significant benefit.

Notes

(1) You should be able to compile a directory in your radio of frequently used MMSI numbers, in which case just select the one you want from the list.

(2) Remember to listen to the relevant working channel before initiating the DSC call to make sure it is not busy.

(3) Your radio will probably give you a list of suitable channels to select from. If not, you can enter one.

8

Routine Calls by Voice

Most of the calls you make with your VHF radio will be Routine rather than, I hope, anything more serious. Unlike Distress or Urgency messages, you don't need to follow a set format, but you do need to follow the right procedures and use recognised terminology. If you get it right, you will sound professional, competent and unambiguous.

In all the following examples, transmissions made by you in yacht *Voyager* are shown in **bold**, by the yacht *Seabird* in regular, and by a third party (coastguard, marina, etc.) in **heavy**.

Set up the radio

- Turn on the power.

- Adjust the squelch. To do this, turn the knob until you hear a loud 'hiss', then turn it back until it *just* disappears.

- Turn the volume to a comfortable level. This is best done while you are adjusting the squelch.

- Select Ch 16.

> You may wish to include normal civilities (please, thank you, etc.) during Routine calls, but be wary of sounding too casual and allowing the conversation to slip into an informal 'phone call'.

- Choose high or low power, depending on the range of the other vessel.

- Hold the microphone a few centimetres from your mouth and slightly to one side, press the PTT switch and speak normally and clearly.

Having got everything ready, it's time to transmit.

Calling another yacht

Initial call on Ch 16

> **Seabird, Seabird this is Voyager, Voyager. Over** [1]

Seabird replies and designates a working channel [2]

> **Voyager this is Seabird, Channel six. Over**

You confirm the working channel

> **This is Voyager. Channel six. Over**

Both vessels switch to Ch 6, *Seabird* replies

> **Voyager this is Seabird**... [and continue the conversation]

> Think about what you want to say – perhaps jot it down – before picking up the microphone. This is always recommended for calls which require specific information, such as informing the coastguard of your passage plans.

(1) For the initial call, it may be necessary to repeat the name of the vessel you are calling and/or your name, but no more than three times. In quiet conditions or when calling a station on a designated channel (eg Southampton VTS on Ch 12), once should be ample:

> Southampton VTS this is yacht *Voyager*. Over

(2) When changing to a working channel, it should be the vessel *being called* (in our example, *Seabird*) who decides which channel to use. However, if you have already checked that a particular working channel is clear, or if *Seabird* does not suggest a channel, it may be quicker for you to nominate that channel after *Seabird* has acknowledged the initial call.

Initial call on Ch 16 (*Voyager* nominates the working channel)

> **Seabird, Seabird** this is **Voyager, Voyager. Over**

Seabird replies

> **Voyager** this is *Seabird*. Over

Voyager nominates the working channel

> **Seabird** this is **Voyager. Channel six. Over**

Seabird confirms the working channel

> **Voyager** this is *Seabird*. Channel six. Over

Both vessels switch to Ch 6, *Voyager* replies

> **Seabird** this is **Voyager...** [and continues the conversation]

Calling the coastguard with a passage report

Initial call on Ch 16

> **Falmouth Coastguard this is yacht *Voyager*, *Voyager*. Routine traffic. Over** [1] [2]

Falmouth CG replies and designates a working channel

> **Voyager this is Falmouth Coastguard, channel six seven. Over** [3] [4]

You confirm the working channel

> **This is *Voyager*, channel six seven. Over**

Both stations switch to Ch 67, Falmouth Coastguard replies on Ch 67

> **Voyager this is Falmouth Coastguard, pass your traffic. Over**

The following should be included in your reply

> **Falmouth Coastguard this is *Voyager***
> **Callsign: Mike Whisky November Five**
> **MMSI: 235 123 456**
> **I departed Dartmouth at 0700 BST** [5]

on passage to St Peter Port

ETA St Peter Port is 1800 BST

Two adults and one child on board

Over

Falmouth CG replies

***Voyager*. All received, have a good trip. Out** [6] [7]

Notes

(1) Indicate the type of your vessel ('yacht') if it might help.

(2) When calling the coastguard, say if your traffic is Routine, otherwise they may assume it is Safety traffic. The subsequent clarification just wastes time.

(3) Some coastguard stations prefer you to call directly on a working channel (Ch 67 for Solent CG). Check in *Reeds*.

(4) In busy periods, you may be asked to 'stand by' on a working channel. In which case your reply would be:

This is *Voyager*. Channel six seven. Standing by. Over

(5) To avoid confusion, always include the time zone. If you are sailing from the UK to France, for example, you should use the time kept in the country of the station you are calling (in this case, BST) for both your ETD and ETA. Or use GMT and remember to adjust for the time zone you are keeping.

(6) '*Out*' so no need for a reply!

(7) UK coastguard stations will often ask you to inform them of your safe arrival. Don't worry too much if this is difficult, or you forget, as no action will be taken unless the coastguard believes that you might be in trouble. For this reason, always ask someone ashore to be your 'shore contact', and be sure to keep them informed of your plans. Let them know when you arrive safely, or when to raise the alarm if they haven't heard from you by a certain time.

Calling a marina for a berth

This is an example of where a slightly more relaxed style may be appropriate.

Initial call on Ch 80 [1]

Mayflower Marina this is yacht *Voyager*. Over

Reply by the marina

Voyager this is Mayflower Marina. Good afternoon, how may I help? Over

Voyager

This is *Voyager*, [2] good afternoon. I would like a berth for one night, please. My overall length is 10.6 metres, draft 1.8 metres. Over

[Continue until you have been allocated a berth]

(1) Usually Ch 80, but may be another channel. Check before calling.

(2) Once you have established contact, there is no need to repeat the name of the station you are calling (the marina, in this example), but it is prudent to include your name before each transmission to avoid any confusion – your berth might be allocated to another boat by mistake.

Radio check

I have already made my views known about the necessity of conducting radio checks (see Chapter 6). If you think one is essential, *avoid using Ch 16* and try to call a station other than the coastguard, which will invariably have more pressing matters to attend to. In rough order of preference conduct your check with:

- Another yacht on a pre-arranged working channel. You could organise this either before setting out or make initial contact by DSC.
- Your local National Coastwatch Institution (NCI) station on Ch 65.
- Your marina or yacht/sailing club, probably on Ch 80.
- Your own handheld VHF set. This may prove your main radio is transmitting but will not be a good test of power output.
- The coastguard on a designated working channel.
- The coastguard on Ch 16 – last resort.

Initial call on Ch 65

Shoreham NCI this is yacht *Voyager*. Radio check, please. Over

Reply from the NCI

***Voyager*, you are loud and clear on Channel 65. Over**

End call

This is *Voyager*, thank you. Out

 If you are '*weak and distorted*' (or almost anything other than '*loud and clear*' or '*good and readable*'), you may have a problem. Don't ignore it; try another check with an alternative station or get your system checked out.

9
Distress calls – MAYDAY

D efinition of Distress:

A vessel or person is in grave and imminent danger and requires immediate assistance.

The important words here are:

Vessel or person You may not agree, but a dog overboard does **not** warrant a Distress call.

Grave Risk to life or of **serious** injury to a person or of major damage or loss of the vessel.

Imminent There is no time to lose; you need assistance **as soon as possible.**

If your predicament does not meet *all* these criteria, **do not** send a Distress alert; send an Urgency announcement instead (see Chapter 10). False or needless Distress alerts can unnecessarily mobilise rescue assets – lifeboats, helicopters, aeroplanes – which will not then be available for a genuine emergency.

An Urgency announcement can always be upgraded to Distress if the situation deteriorates.

If you have a DSC radio, send the initial alert via DSC. This is swifter and avoids the risk of misunderstandings and consequent delays. You will need to follow this up with a voice call.

At Appendices A and B you will find an example of a check-off list of actions to be taken and a pro forma for a voice Distress call. Both may be copied and adapted for your vessel. Once made, it should be permanently displayed next to the radio.

When making a Distress call, you really must try to stick rigidly to the correct procedure in order to get your message through unequivocally and fast. Easier said than done when the water is up to your ankles and rising, I know, but the quicker your plea for help is received and understood, the faster the organisation can swing into action to provide the appropriate assistance.

Position

Before you make your call, you must know where you are. This may sound obvious, but if you don't have access to your latitude and longitude from GPS you will need to measure it from the chart or be ready to describe your position. *Write it down.* Without a position, the rescue services may not be able find you, although they may be able get a very rough position by intercepting your VHF signals with direction-finding equipment.

1. Latitude and longitude

If your radio has a GPS input, your position will be continuously updated. If not, you will have to enter it manually. This is a lengthy and error-prone job but is essential when using the DSC function for a Distress alert. It should be done at regular intervals, ideally hourly.

The preferred and most accurate way to define your position is by lat & long. Take, as an example:

50° 32'·235N 001° 45'·155W

Sent by voice you would say:

Five zero degrees, three two decimal two three five minutes north; zero zero one degrees, four five decimal one five five minutes west

To save time and possible errors, you only need to express the minutes to one decimal place which, when rounded to the nearest minute, will be within 100 metres of your actual position – perfectly adequate for most purposes. For more accuracy, two decimal places will place you within 10 metres. Also, although degrees of longitude are usually defined by three digits, any initial zeros may be omitted if by doing so you would not cause any confusion. Our spoken position then becomes rather shorter and, consequently, easier to write down:

Five zero degrees, three two decimal two minutes north, one degree four five decimal two minutes west

2. Range and bearing

In the absence of a lat & long, the next best option is your range and bearing *from* a well-known charted mark. The position above is:

220° from the Needles lighthouse at a range of 9.5 miles

By voice this would be:

Two two zero degrees, nine decimal five miles from the Needles lighthouse

or

Two two zero degrees from the Needles lighthouse, range nine decimal five miles

3. General area

Finally, if time does not allow for anything more accurate, a general description of where you are is better than nothing. In the example above, this would be:

About 10 miles south west of the Needles

Other details

As well as your position, you will need the following information to hand:

- Name of your vessel
- Callsign (could be omitted if time is short)
- MMSI
- Type of emergency (flood, fire, grounding, collision, etc.)
- The number of people on board (adults and children)
- What assistance is required
- Any other relevant information – injuries, etc.

Initial alert by DSC

All DSC VHF radios have a red Distress button. To operate it, and to avoid false alerts, you will need to slide a cover or lift a flap before pressing the button. The typical routine is: [1]

1. Press the red Distress button and hold for five seconds until it bleeps. [2]

Or, if time allows:

1. Press the red Distress button briefly, once, then select the type of emergency from the menu. [3]
2. Press and hold red Distress button until it bleeps.

 ## Notes

(1) All radios are different. It is *essential* that you read the manual and thoroughly familiarise yourself and your crew with the routine for sending a Distress alert.

(2) This will initiate an 'undesignated' alert.

(3) Selecting the nature of your situation takes very little extra time and forewarns the rescue services. This is far better than an undesignated alert, particularly if you are then unable to send any follow-up messages. The options are: *Fire, Flood, Collision, Grounding, Capsizing, Sinking, Adrift, Abandoning, Piracy and Man overboard.*

Follow-up voice alert

Once you have sent a DSC Distress alert, the radio will continue sending it at regular intervals until someone acknowledges. If no acknowledgement is received within a few seconds, you should send a voice alert. Even if someone does respond, you will almost certainly be asked to repeat the alert, so be ready.

Make sure Ch 16 and high power (25W) are selected; speak slowly and clearly.

The format to be followed is:

MAYDAY, MAYDAY, MAYDAY [1]
This is yacht *Voyager*, *Voyager*, *Voyager* [1]
MMSI 235 123 456 [2]
Callsign Mike Whisky November Five [3]
MAYDAY Voyager
My position is [see above]
My vessel is on fire, which is out of control [4]
There are two adults and one child on board
I require immediate assistance
Over [5]

Notes

(1) Follow this exact format, and repeat 'MAYDAY' and the vessel's name three times at this stage of the alert.

(2) Your MMSI should be included so that your identity can be positively linked to your DSC alert (if one has been sent).

(3) The callsign may help with identity but could be left out if time is short.

(4) More details of the emergency can be passed later. The priority is to get the alert sent as quickly as possible.

(5) At this point, release the transmit button and listen for a response. If nothing is heard after a minute or so, repeat the call.

Cancelling a Distress alert

False DSC Distress alerts are, thankfully, very rare but they do happen. If you send one in error, you must cancel it as soon as possible.

🔋 Turn off the radio [1] (or DSC controller if separate) to prevent automatic repetition of the alert.

🔋 Switch the radio back on, select Ch 16 and high power (25W).

🔋 Say your message.

> **All stations, All stations, All stations**
> **This is yacht *Voyager*, *Voyager*, *Voyager***
> **MMSI 235 123 456**
> **Callsign Mike Whisky November Five**
> **Cancel my Distress alert** [2]
> **I say again, cancel my Distress alert**
> **Out**

(1) Some radios include a function to cancel a Distress alert without having to turn off and back on again. Check the handbook.

(2) Give the time of the false alert if you know it.

Responding to a Distress alert

In accordance with the International Convention for the Safety of Life at Sea (SOLAS), we are all under an obligation to respond to Distress signals and to render assistance if possible. In a yacht or other small craft this may not be practicable, in which case wait five minutes to see if a more suitable station responds. If you don't hear anything, you should respond:

Mayday *Snowgoose*
This is yacht *Voyager*, *Voyager*, *Voyager*
Received Mayday
Out

If you are able to render assistance, call *Snowgoose* again as soon as possible with your intentions, including your position and your ETA with her. Otherwise, send a MAYDAY RELAY (see below).

 If the alert has been received by DSC follow the instructions for your radio. You will have the choice to *Accept* the MAYDAY or to *Quit*.

MAYDAY RELAY

If you know, or suspect, that another vessel is in distress, you are bound to take action yourself. You may have seen red flares or other Distress signals, or you may have heard a weak Distress call which has not apparently been acknowledged. In this situation, wait five minutes then transmit a 'MAYDAY RELAY' by voice.

An initial Urgency announcement by DSC or voice is not necessary and should not be sent.

The main thing is to make it quite clear that it is not you who is in distress, hence the format for MAYDAY RELAY:

MAYDAY RELAY, MAYDAY RELAY, MAYDAY RELAY
Falmouth Coastguard, Falmouth Coastguard, Falmouth Coastguard [1]
This is yacht *Voyager*, *Voyager*, *Voyager*
MMSI: 235 123 456
Received the following **MAYDAY** from yacht *Snowgoose*: [2]
[Repeat the Distress call you heard, or describe the situation]
Over

 (1) If you know you are within range, call the nearest coastguard station (as above). If there is no response, send the MAYDAY RELAY to 'All Stations'.

(2) If you have seen or heard a Distress situation but not heard an actual MAYDAY call, simply say what you have seen or heard. As always, keep it short and be prepared to give more details once you are in contact with another vessel or the coastguard.

Controlling Distress communications

Distress communications on Ch 16 automatically impose general radio silence until the incident is over. This is one reason why you must listen before transmitting to be sure you are not about to interrupt urgent traffic. If necessary, radio silence can be enforced or lifted by the station (eg the coastguard) controlling the incident. This is done by:

Seelonce Mayday – all traffic unconnected with the emergency is prohibited.

Seelonce Feenee – lifts radio silence

If you are involved in a Distress situation you must continue monitoring the relevant working channel until you are formally released by the coordinating authority, normally the coastguard.

10
Urgency Announcements — PAN PAN

The definition of Urgency is:
A vessel or station has a **very urgent message concerning the safety of a ship or person.**

An Urgency announcement should be used where there is no *immediate* danger to a person or vessel, and *immediate* assistance is not necessary. If the situation worsens, a PAN PAN call may be upgraded to a MAYDAY.

 Urgency, Safety and Routine calls are 'announcements'. The term 'alert' is only used for Distress calls.

Urgency announcement by DSC

Unlike a Distress alert, an Urgency announcement by DSC does not include your position. Otherwise, the format is very similar with some variation in terminology.

 You may wonder why a DSC call is needed if you have to send a follow-up voice message anyhow. The reason is that it will alert other stations by sounding an alarm (as does a Distress alert) on all receiving radios.

Select the DSC option on your radio, then select 'Urgency'.

Select the channel – normally Ch 16.

Press 'Send' (or whatever your radio requires to transmit the call).

Wait for a response, then send your message by voice.

PAN PAN, PAN PAN, PAN PAN
All Ships, All Ships, All Ships
This is yacht *Voyager*, *Voyager*, *Voyager*
MMSI 235 123 456
My position is [lat & long, range & bearing or general description]
I have been dismasted and am drifting towards the Shingles Bank
I require a tow into the Solent
Two adults on board
Over

Retransmit your message if you get no reply. How often you do this will depend on the situation. Don't hesitate to make a MAYDAY call if that becomes necessary.

Responding to a PAN PAN call

Follow the same routine as for a MAYDAY call. Urgency DSC alarms are not unusual, and many originate from distant shore stations. More often than not the only action required is to cancel the alarm on your radio. However, always listen carefully to a spoken PAN PAN message just in case you can help.

Urgent medical advice

Make a PAN PAN call if you need urgent medical advice unless the casualty's condition is life-threatening, in which case a MAYDAY call is appropriate. The coastguard will give you a working channel and then put you in contact with a doctor. There is no charge for this.

 The old proword 'PAN PAN Medico' is no longer used.

11

Safety Announcements – SÉCURITÉ

A Safety call is used to broadcast important navigational or meteorological information which might be of interest to ships in the area. Situations when it might be appropriate for you to make a *SÉCURITÉ* (pronounced 'say-cure-ee-tay') call include:

- The sighting of a semi-submerged object which poses a danger to small craft.

- You see a navigational buoy which is clearly adrift.

- You are becalmed without an engine in a busy shipping lane and wish to warn other vessels of your position.

The list is not exhaustive but gives you an idea of when a Safety message might be required.

By DSC and Voice

To alert other stations, select the *All Ships* and *Safety* option on the DSC menu and then *Transmit*. Follow this by a voice message on the selected channel – normally Ch 16:

> SÉCURITÉ, SÉCURITÉ, SÉCURITÉ
>
> All ships, All ships, All ships
>
> This is yacht *Voyager*, *Voyager*, *Voyager*
>
> MMSI 235 123 456
>
> My position is ...
>
> Probable shipping container adrift and semi-submerged in position ...
>
> Over

Listen for a response and be prepared to provide more details if requested.

> You will hear many Safety announcements. French coast stations often precede MSI broadcasts with '*SÉCURITÉ, SÉCURITÉ, SÉCURITÉ*'; the UK coastguard generally does not.

12
The Phonetic Alphabet

Y ou need to know the phonetic alphabet, not only to pass the SRC exam but also to avoid embarrassing hesitations while you rack your brain for the correct word as you spell your name. 'S for sugar, E for elephant ...' doesn't really cut the mustard.

If you need to spell your boat's name or callsign (as you often have to), always do so using the phonetic alphabet. It is carefully designed to be unambiguous, even to those whose first language is not English.

Phonetic alphabet			
A	Alfa	N	November
B	Bravo	O	Oscar
C	Charlie	P	Papa
D	Delta	Q	Quebec
E	Echo	R	Romeo
F	Foxtrot	S	Sierra
G	Golf	T	Tango
H	Hotel	U	Uniform
I	India	V	Victor
J	Juliet	W	Whisky
K	Kilo	X	X-ray
L	Lima	Y	Yankee
M	Mike	Z	Zulu

There are occasions when you should use 'normal' spelling rather than the phonetic alphabet. When calling Southampton VTS, it would be a nonsense to say 'Southampton Victor Tango Sierra'. Similarly, don't be tempted to spell out the name of a well-known object if it is usually referred to by its initials. The Outer Spit Buoy at the entrance to Portsmouth Harbour is a good example. Say either, 'Outer Spit Buoy' or 'OSB', not 'Oscar Sierra Bravo'.

Numbers are generally pronounced normally but sometimes with a slight change in emphasis.

Numerals			
1	Wun	6	Six
2	Too	7	Seven
3	Tree	8	Ait
4	Fower	9	Niner
5	Fife	0	Zero

If you are naming your vessel for the first time, or if you are changing her name (no, it's not unlucky!), think what it will sound like over the radio in an emergency, and how long it will take to spell phonetically. *It'stoowindy* (a real name) is a disquieting case in point.

Also, avoid choosing a name which could be misinterpreted. *Heyday* and *Can Can* are obvious examples.

13
Glossary

M ost of the terms you will come across while using your VHF radio have already been covered. They are listed here for ready reference, plus a few more.

AIS Automatic Identification System.

ALRS Admiralty List of Radio Signals. Published in six volumes and covers all you want to know about maritime communications – and considerably more.

ATIS Automatic Transmitter Identification System. For use in European inland waterways. See RAINWAT opposite.

Broadcast Sending a message without specifying who it is intended for.

BST British Summer Time, also known as daylight saving time. BST is one hour ahead of GMT (see below) from the end of March to the end of October (actual dates vary year by year).

CGOC Coastguard Operations Centre. Formerly known as MRCC (see below).

Coast radio station Any maritime radio station ashore, including coastguard stations.

CROSS Centre Régional Opérationnel de Surveillance et Sauvetage (French equivalent of a UK CGOC).

Dual-watch The facility on a radio to monitor two channels at the same time. Usually Ch 16 and one other, set by the operator.

Duplex A channel which transmits on one frequency and receives on another.

EPIRB Emergency Position Indicating Radio Beacon.

ETA Estimated Time of Arrival

ETD Estimated Time of Departure

FM Frequency Modulation.

GMT Greenwich Mean Time. The standard time, sometimes called legal time, in the UK. (See BST above and UTC below).

GPS Global Positioning System.

IMO International Maritime Organisation.

INMARSAT Inmarsat plc, a UK company. Originally the International Mobile Satellite Organisation.

ITU International Telecommunication Union. An agency of the United Nations responsible for issues concerning information and communication technologies.

LW Long Wave.

MCA or MCGA Maritime and Coastguard Agency.

MMSI Maritime Mobile Service Identity.

MRCC Maritime Rescue Coordination Centre. The old term for Coastguard Operations Centre (CGOC).

MSI Maritime Safety Information.

NAVAREA Geographic areas for which governments are responsible for navigation and weather warnings. The UK government is responsible for Navarea 1.

Navtex Navigational **T**el**ex**. For navigation and weather warnings and forecasts.

NMEA National Maritime Electronics Association. The standard interface for connecting electronic devices such as GPS, radar and chart plotters.

NMOC National Maritime Operations Centre (near Fareham, Hampshire).

Ofcom UK's Office of Communications.

PLB Personal Locator Beacon.

PTT Press to Transmit [button].

RAINWAT Regional Arrangement on Telecommunication Service on Inland Waterways. (Applicable on the continent; the UK is not a signatory.)

RIB Rigid Inflatable Boat

RX Abbreviated form of Receive (see TX below).

RYA Royal Yachting Association

SAR Search and Rescue.

SART Search and Rescue Transponder

Scan Setting on a VHF radio to monitor several, or all, channels. Can be set to scan all selected channels in turn ('memory scan') or to monitor Ch 16 between each selected channel ('priority scan').

Semi-duplex The radio uses one aerial to switch between two frequencies on one channel.

Simplex A channel which transmits and receives on the same frequency.

SOLAS Safety of Life at Sea [Convention].

Squelch Suppresses background interference ('hiss').

Tri watch Like dual watch but monitoring Ch 16 and two other channels.

TX Abbreviated form of Transmit or Transmission (see RX above)

UKCG UK Coastguard.

UTC/UT Coordinated Universal Time/Universal Time. For all practical purposes, UTC is the same as GMT.

VHF Very High Frequency

VTS Vessel Traffic Service.

14
VHF Channels and Their Uses

This is a simplified table of VHF channels available for use at sea. Your VHF radio manual will include the frequencies of each channel. You don't need to know them, so I have not included them.

Notes

Ship-to-Ship channels

Also known as inter-ship channels.

The preferred channels are shown in **bold**.

Port Operations

Avoid using these channels for ship-to-ship communications unless specifically designated.

Public Correspondence

These are international channels for making calls over the public telephone network – sometimes referred to as 'link calls'.

Channel Number	Main Uses	Simplex (S) or Duplex (D)	Notes
0	Search and Rescue	**S**	SAR organisations only
1	Public Correspondence, Port Operations	**D**	
2	Public Correspondence, Port Operations	**D**	
3	Public Correspondence, Port Operations	**D**	
4	Public Correspondence, Port Operations	**D**	

Channel Number	Main Uses	Simplex (S) or Duplex (D)	Notes
5	Public Correspondence, Port Operations	D	
6	**Ship-to-Ship**	S	
7	Public Correspondence, Port Operations	D	
8	**Ship-to-Ship**	S	
9	Ship-to-Ship, Port Operations	S	
10	Ship-to-Ship, Port Operations	S	Some MSI broadcasts
11	Port Operations	S	
12	Port Operations	S	
13	Bridge-to-Bridge	S	International Navigation Safety
14	Port Operations	S	
15	Ship-to-Ship, Port Operations	S	Limited to 1W
16	*Distress, Safety and Calling*	S	
17	Ship-to-Ship, Port Operations	S	Limited to 1W
18	Public Correspondence, Port Operations	D	
19	Public Correspondence, Port Operations	D	
20	Public Correspondence, Port Operations	D	
21	Public Correspondence, Port Operations	D	
22	Public Correspondence, Port Operations	D	
23	Public Correspondence, Port Operations	D	
24	Public Correspondence, Port Operations	D	

Channel Number	Main Uses	Simplex (S) or Duplex (D)	Notes
25	Public Correspondence, Port Operations	D	
26	Public Correspondence, Port Operations	D	
27	Public Correspondence, Port Operations	D	
28	Public Correspondence, Port Operations	D	
60	Public Correspondence, Port Operations	D	
61	Public Correspondence, Port Operations	D	
62	Public Correspondence, Port Operations	D	UK MSI broadcasts
63	Public Correspondence, Port Operations	D	UK MSI broadcasts
64	Public Correspondence, Port Operations	D	UK MSI broadcasts
65	National Coastwatch Institution	D	In the UK, dedicated channel for the NCI
66	Public Correspondence, Port Operations	D	
67	UK Small Ship Safety	S	Calling channel for some coastguard stations
68	Port Operations	S	
69	Ship-to-Ship, Port Operations	S	
70	*Digital Selective Calling*	S	DSC **only**
71	Port Operations	S	

Channel Number	Main Uses	Simplex (S) or Duplex (D)	Notes
72	Ship-to-Ship	S	
73	Ship-to-Ship, Port Operations	S	
74	Port Operations	S	
75	Ship-to-Ship	S	
76	Ship-to-Ship	S	
77	Ship-to-Ship	S	
78	Public Correspondence, Port Operations	D	
79	Public Correspondence, Port Operations	D	
80	Marinas	D	UK only
81	Public Correspondence, Port Operations	D	
82	Public Correspondence, Port Operations	D	
83	Public Correspondence, Port Operations	D	
84	Public Correspondence, Port Operations	D	
85	Public Correspondence, Port Operations	D	
86	Public Correspondence, Port Operations	D	
87	Port Operations	S	
88	Port Operations	S	
M1 (P1)	Marinas, Yacht Clubs	S	Private channel, UK only
M2 (P2)	Race Control, etc.	S	Private channel, UK only

15

SRC Syllabus and Assessment

To obtain your Short Range Certificate, you will be expected to demonstrate your knowledge of the following, either during the practical test or the written exam. I have included references to the relevant chapters in this book.

In the next chapter you will find some questions and answers to test your knowledge and understanding.

Distress situations

Define a Distress situation (Chapter 9)
Initiate a Distress alert, input position and time manually or
    automatically (Chapter 9)
Send a MAYDAY message by voice (Chapter 9)
Respond to a DSC Distress alert and MAYDAY message (Chapter 9)
Send a MAYDAY Relay message (Chapter 9)
Know how to deploy an EPIRB and a SART (Chapter 2)
Know the procedure for cancelling a Distress message sent in
    error (Chapter 9)

Urgency situations

Identify a situation where an Urgency message is appropriate
    (Chapter 10)
Initiate a DSC Urgency announcement (Chapter 10)
Send a PAN PAN message by voice (Chapter 10)
Respond to an Urgency message (Chapter 10)

Safety situations

Identify a situation where a Safety message is appropriate (Chapter 11)
Initiate a DSC Safety announcement (Chapter 11)
Send a Safety message by voice (Chapter 11)
Receive Maritime Safety Information (MSI) by Navtex (Chapter 2)

Routine communications

Initiate a Routine DSC call (Chapter 7)

Establish communications and exchange messages by voice, using
appropriate channels (Chapters 4, 6 & 7)

Make a test call (radio check) (Chapter 6)

Enter a Group and Individual MMSI (Chapter 2)

Maintain appropriate listening watch on DSC and voice channels
(Chapter 5 – VHF channels, Chapter 6 – Radio procedures)

Use the phonetic alphabet (Chapter 12)

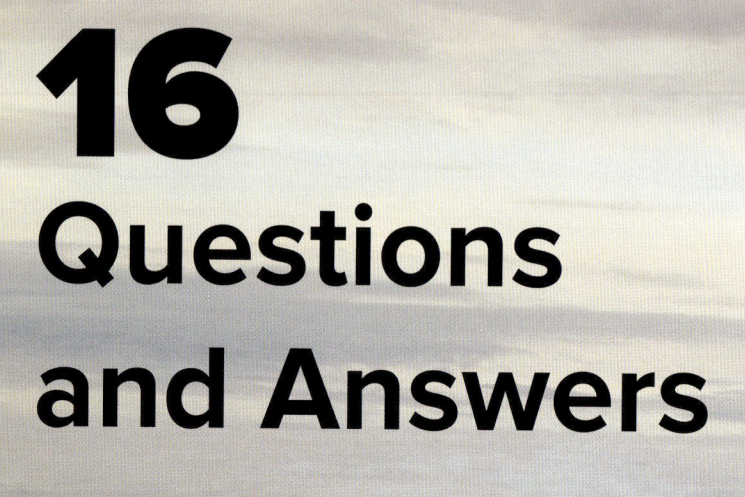

16
Questions and Answers

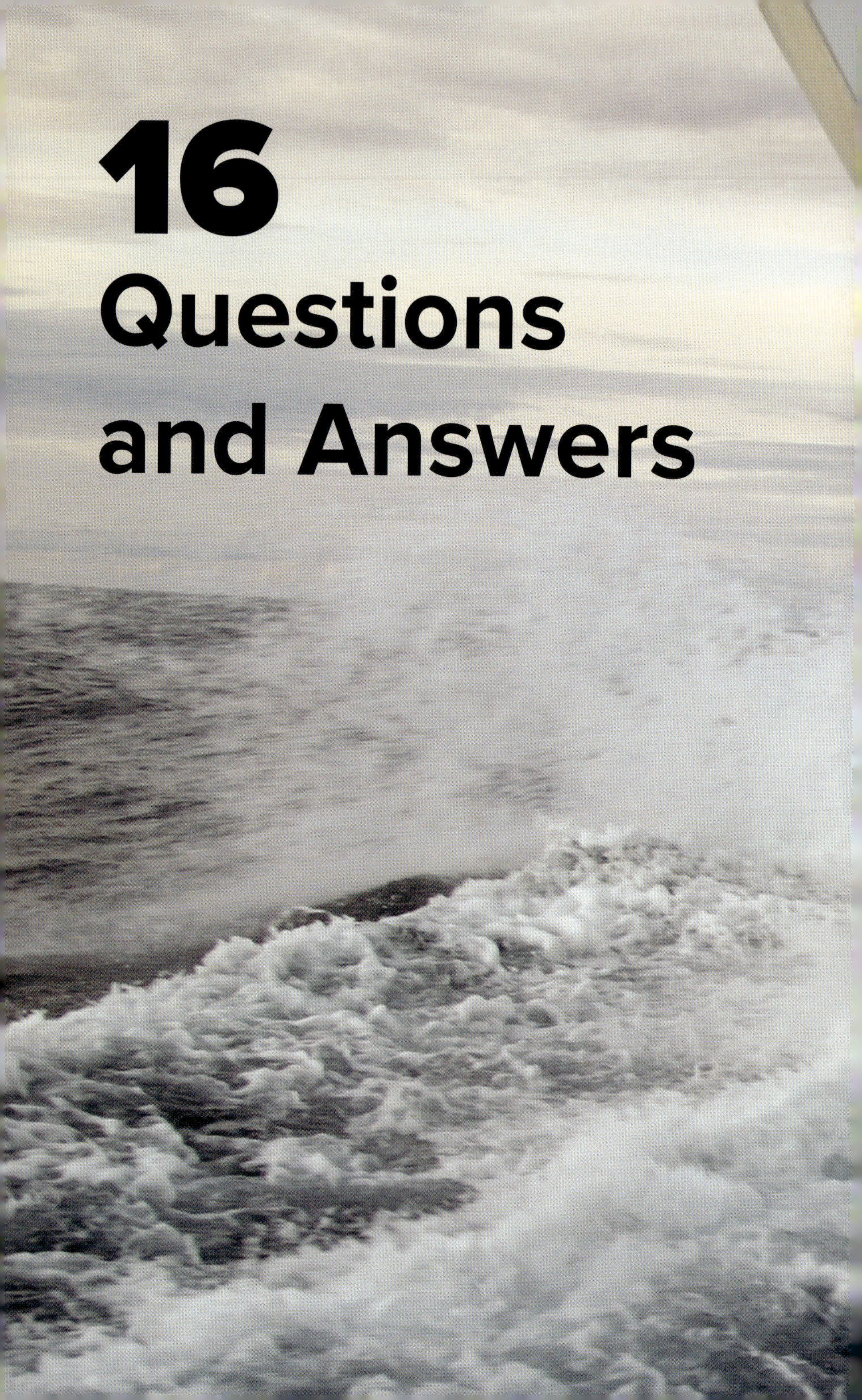

These are not the exact questions you will be asked during your SRC exam, but if you can answer all of them correctly you should have no trouble passing. Some go slightly beyond the syllabus, but the answers should be known, I suggest, by any reasonably knowledgeable and proficient VHF radio operator.

The questions are generally grouped in the same order as the previous chapters. Like the SRC written exam, many are multiple choice, but if you are not sure of an answer don't guess and hope for the best. Go back to the relevant chapter to check your understanding.

In multiple-choice questions select all answers that apply.

Radio theory and GMDSS

1.   Which frequency band is used for short range maritime communications?
  a) High frequency (3 – 30kHz)
  b) Very high frequency (30 – 300MHz)
  c) Super high frequency (3 – 30GHz)

2.   Low frequency (30 – 300kHz) is used for transmitting what essential information to mariners?
  a) The times of high water at Dover
  b) Shipping forecasts
  c) Inshore waters forecasts
  d) Vessel Traffic Services

3.   Why do VHF transmissions have a limited range?
  a) They don't 'bend' over the horizon
  b) They are limited to 'line of sight'
  c) Their typical power output limits them to just a few miles
  d) Range can be increased by shouting

4. Using a **VHF** radio on high power (**25W**) in a sailing yacht with an aerial height of about **15** metres, what is the maximum range you would expect to be able to talk with a similarly fitted vessel?
   a) 5 – 10 miles
   b) 10 – 15 miles
   c) 15 – 20 miles
   d) 20+ miles

5. Using a **VHF** radio on high power (**25W**) in a motor cruiser with an aerial height of **5** metres, what is the maximum range would you expect to be able to talk with a merchant ship (aerial height **50** metres)?
   a) 10 miles
   b) 15 miles
   c) 20 miles

6. Why should you use low power (**1W**) on your **VHF** radio whenever possible?
   a) To prevent the radio overheating
   b) To limit range so you can have a long chat without others overhearing you
   c) To avoid interfering with other traffic on the same frequency

7. What does **GMDSS** stand for?
   a) Global Marine Digital Safety Service
   b) Greater Manchester Daily Statistics System
   c) Global Maritime Distress & Safety System
   d) Good Marmalade Doesn't Spread Smoothly

8. Which GMDSS Sea Area covers the coastal waters of the UK?
   a) A1
   b) A2
   c) A3
   d) None of the above

9. What is the definition of GMDSS Sea Area A1?

10. When on passage across the North Sea from the Humber to Denmark, which GMDSS Sea Area(s) would you be in?
    a) A1 and A2
    b) A1, A2 and A3
    c) A2 and A3

11. On the same passage across the North Sea (Q10) would you *need* an MF radio for safety purposes? Explain your answer.

12. GMDSS comprises which of the following?
    a) VHF DSC radio
    b) MF radio
    c) Satellite systems
    d) Navtex
    e) EPIRBs, PLBs and SARTs
    f) All of the above

13. What is the purpose of the Cospas/Sarsat satellite system?

14. What frequency is used by Cospas/Sarsat to receive signals from distress beacons?
    a) 406MHz
    b) 121.5MHz
    c) 243MHz
    d) All of the above

15. What is the main purpose of Navtex?

16. What language is used for Navtex transmissions?
    a) French
    b) English
    c) Mandarin

17. What is the maximum range you would expect to be able to receive Navtex broadcasts?
    a) 100 miles
    b) 200 miles
    c) 400 miles

18. Where could you find a Navtex message identity (eg GE 05)?

19. What do the following stand for?
    a) EPIRB
    b) PLB
    c) SART

20. What is the difference between a Cat 1 and a Cat 2 EPIRB?

21. With whom should you register your EPIRB?

22. What should you do if you activate your EPIRB accidentally?

23. What does a SART do?

    a) Sends a signal to the Search and Rescue (SAR) services

    b) Communicates with satellites

    c) Shows up as a series of dots on the radars of receiving vessels

DSC, AIS and ATIS

24. What does MMSI stand for?

    a) Mobile Marine Situation Indicator

    b) Mobile Marine Status Instruction

    c) Maritime Mobile Service Identity

25. Who supplies the MMSI for your radio?

    a) Ofcom

    b) Maritime and Coastguard Agency

    c) RYA

26. Why should you be particularly careful when entering the MMSI for the first time?

27. To what is an MMSI starting with '00' allocated?

    a) A group of yachts

    b) A British warship

    c) A coast station

28. Is an MMSI for a handheld radio allocated to the parent vessel or the radio?

29. On which VHF channel does DSC transmit?

    a) Ch 16

    b) Ch 67

    c) Ch 70

30. Can you use the specified DSC channel for voice communication in an emergency?

31. When should you *not* transmit by DSC?

    a) When more than 50 miles offshore

    b) When on inland waterways

    c) For Routine calls

32. Is AIS an integral part of GMDSS?

33. What is the primary function of AIS in a yacht?

    a) Collision avoidance

    b) Navigating in fog

    c) A substitute for radar

34. What does ATIS stand for?

    a) Automatic Tracking of Inland Ships

    b) Automated Traffic Indicating Service

    c) Automatic Transmitter Identification System

35. How does an ATIS MMSI differ from your vessel's MMSI?

VHF radios

36. How should you use the squelch control?

    a) Turn to get maximum 'hiss'

    b) Turn until you just cut out any 'hiss'

    c) Don't bother, it's too difficult

37. What is the PTT switch?

    a) Push To Talk

    b) Practice Test Transmission

    c) Press to Transmit

38. Why is the red Distress button covered?

39. **How do you select High Power?**
    a) Press the Ch 16 button
    b) Press the Hi/Lo button
    c) Either of the above

40. **Why should a VHF aerial be sited as high as possible?**
    a) To keep it away from possible damage
    b) To prevent radiation of the crew
    c) To maximise range

41. **When should you *not* use the DSC function on a handheld radio?**
    a) Within 3 miles of the coast
    b) On UK inland waterways
    c) Outside UK territorial waters

Rules and regulations

42. **What 'authority' do you need to install and use a fixed VHF radio in your boat?**
    a) Nothing
    b) Ship Radio Licence
    c) Short Range Certificate
    d) Ship Portable Radio Licence
    e) Radio log

43. **When should you inform Ofcom about changes to your Ship Radio Licence?**
    a) Change of ownership of the boat
    b) Addition of a radar
    c) Every five years

44. What equipment, if installed, must be included on your Ship Radio Licence?
    a) MF and HF radios
    b) AIS transponders
    c) Radar
    d) Handheld VHF radio
    e) EPIRB
    f) SART
    g) None of the above
    h) All of the above

45. Do you need to register a PLB?

46. What is your authority to operate your VHF radio?
    a) Ship Radio Licence
    b) No licence needed
    c) RYA Day Skipper qualification or above
    d) Short Range Certificate

47. Who issues a Short Range Certificate?
    a) Ofcom
    b) Maritime and Coastguard Agency
    c) RYA
    d) Authorised yacht club

48. Are you required to maintain a log of calls made and received on your VHF radio?

49. What is the minimum age for taking the SRC exam?
    a) No minimum age
    b) 14
    c) 16
    d) 18

50. Under what circumstances is someone who does not hold an SRC allowed to use the VHF radio?
    a) To make a PAN PAN call
    b) If an SRC holder is on board
    c) Under direct supervision of an SRC holder
    d) If they hold an RYA Day Skipper qualification

51. Which of the following are you permitted to use to identify yourself when transmitting on a VHF radio?
    a) Your name (eg Joe Bloggs)
    b) Vessel's name
    c) MMSI
    d) International callsign
    e) No requirement to identify yourself

52. Are you allowed to send a Distress alert in order to test your radio?
    a) Yes
    b) No
    c) Only if you suspect your radio is faulty

53. If you are involved in Distress traffic, when may you stop monitoring the relevant working channel?
    a) When the situation appears to have been resolved
    b) When you can no longer offer any assistance
    c) When stood down by the coordinating authority

54. If you wish to inform other vessels that you are becalmed in a shipping lane, but not in any immediate danger, you should:
    a) Broadcast 'blind' (ie not addressed to anyone in particular)
    b) Address only those ships you can identify – perhaps on AIS
    c) Use the callsign 'All stations'

Radio procedures

55. Having selected the appropriate channel, what should you always do before transmitting?

   a) Say 'Testing, testing'

   b) Say 'Hello, is anyone there?'

   c) Listen to ensure your chosen channel is not busy

56. What is the main drawback of using high power when calling a nearby vessel?

   a) You may cause damage to the other vessel's radio

   b) With a handheld set the battery will drain more quickly

   c) All radios within about 20 miles may hear your conversation

   d) You may block other, weaker, transmissions

57. When do you *need* to conduct a 'radio check'?

   a) Never

   b) Every time you leave your berth

   c) Before leaving harbour

   d) When you have doubts about your radio's reliability

   e) When you have worked on the aerial run or installed a new radio

58. If you do need to conduct a radio check, put the following options of who to call in order of preference:

   a) Coastguard on Ch 16

   b) NCI station on Ch 65

   c) Another yacht by arrangement

   d) Any station via DSC

   e) A marina or harbour office

   f) Your handheld radio

59. What does 'Over' mean at the end of a transmission?
   a) I am no longer listening on this channel
   b) I have finished my transmission and am waiting for your reply
   c) Did you understand my last transmission?

60. What does 'Out' mean at the end of a transmission?
   a) I am no longer listening on this channel
   b) I am switching off my radio
   c) I do not expect a reply from you

61. When is it appropriate to say 'Over and out' at the end of a transmission?
   a) When you want the other station to reply
   b) To confirm your message has been understood
   c) Never
   d) At the end of a radio check

62. What word should you use to indicate that 'I have received and understood your message'?
   a) Roger
   b) Received
   c) Right-ho
   d) Over and out

63. Are you allowed to call someone ashore using your VHF radio?
   a) Yes
   b) No
   c) Only authorised stations (eg marinas)
   d) Yes, if it is to your handheld set
   e) Yes, if your crew has called you using the handheld set

64. **How should you indicate that you want a repetition of *part* of a message?**
    a) 'Say again. Over'
    b) 'Say again. Out'
    c) 'What?'
    d) 'Please repeat everything you said after [or before] ...'
    e) 'Say again all after [or before] ...'

65. **If you receive a call from a station you cannot identify (eg because of poor reception), how should you reply?**
    a) 'Station calling [my vessel], who are you? Over'
    b) 'Call again, louder. Over'
    c) 'What?'
    d) 'Station calling [my vessel], say again your name. Over'

66. **How long should you wait if you receive no response from a vessel you have called?**
    a) Don't waste time; call again immediately
    b) 1 minute
    c) 2 minutes
    d) 3 minutes

VHF channels

67. **Which are the four preferred channels for ship-to-ship working?**

68. **Which other channels may be used as alternatives to those in question 67 above?**

69. **What is a duplex channel?**

70. **What is a simplex channel?**

71. What is the calling channel for most UK marinas?
    a) Ch 16
    b) Ch 67
    c) Ch 77
    d) Ch 80

72. What is the *main* purpose of Ch 16?
    a) Calling other vessels
    b) Chatting with other vessels
    c) Distress, Urgency and Safety working
    d) Conducting radio checks

73. What is the 'bridge-to-bridge' VHF channel?
    a) Ch 16
    b) Ch 67
    c) Ch 13
    d) Ch 12

74. Which channels are used for MSI broadcasts?

75. Where could you find out which channel is used for MSI broadcasts in your area?
    a) Local newspaper
    b) *Reeds Nautical Almanac*
    c) Admiralty List of Radio Signals
    d) Ofcom website

76. Why can't you hear other yachts calling marinas on Ch 80, but you can hear the marinas' replies?

77. Why should you never use Ch 70 for voice communications?

78. What is Ch 0 used for?

79. What is Ch M1 (often referred to as simply Ch M) used for?

80. Are you allowed to use Ch M abroad?

81. When sailing offshore but close to busy shipping areas (eg southern North Sea), which two VHF Channels would you monitor, and why?

Routine calls by voice

82. Having set up the radio, before transmitting on Ch 16 what are the two most important checks to make?
    a) Listen to Ch 16 to make sure it is not busy
    b) Think what you are going to say
    c) Put your tea somewhere safe

83. How many times should you say the name of the vessel being called?
    a) Once
    b) Twice
    c) Three times
    d) It depends (explain)

84. Who should select the working channel?
    a) You, during the initial call on Ch 16
    b) The other vessel, on receipt of your initial call
    c) Discuss the options when the other vessel replies

85. When making a Routine call to the coastguard to pass your 'passage report', list six items which should be included.

86. **When transmitting, how should you pitch your voice?**

    a) Louder than usual

    b) Softer than usual

    c) Higher pitch than normal

    d) Lower pitch than normal

87. **How should you hold the microphone?**

    a) As close to your mouth as possible

    b) Immediately in front of your mouth

    c) Slightly to one side and about 2 – 5 cm away

    d) At arm's length

88. **What do the rules say about the length of a Routine call?**

Using DSC for Routine calls

89. **From the DSC menu, which option should you select for a Routine call?**

    a) Distress

    b) All ships

    c) Group

    d) Individual

90. **What essential information is needed before making a Routine DSC call to another yacht?**

    a) The other skipper's name

    b) The MMSI number of the other yacht

    c) The other yacht's international callsign

91. If you know the name of a vessel you wish to call by DSC, but not her **MMSI**, what are your options?
    a) Make an All Ships call by DSC
    b) Call by voice on Ch 70
    c) Call by voice on Ch 16

92. What are the main advantages of using DSC for Routine calls rather than voice?
    a) It is quicker
    b) It avoids cluttering up Ch 16
    c) It avoids ambiguity (about who you are calling)
    d) It automatically switches to a working channel

93. Spell the following words using the phonetic alphabet:
    a) Cowes
    b) Dover
    c) Newcastle
    d) Torquay
    e) Lizard Point

Note: test yourself frequently until the phonetic alphabet becomes second nature!

Distress – MAYDAY

94. What are the three vital elements of the definition of Distress?

95. Which of the following would *not* warrant a Distress call? Justify your answers.
    a) Becalmed and without engine 5 miles south of the Needles
    b) Fire on board which is out of control
    c) You have lost sight of your dog which has fallen overboard
    d) Man overboard
    e) Dismasted, no engine and drifting towards rocks 1 mile away
    f) Aground on mud in a river
    g) Crew member struck on the head by the boom and unconscious

96. Which channel should be used for a Distress alert by voice?
    a) Ch 67
    b) Ch 70
    c) Ch 16
    d) Ch 80

97. What information should be included in a Distress message?

98. Give an example of a full Distress call and subsequent message.

99. What are the options for describing your position?

100. What do the following terms mean?
   a) MAYDAY RELAY
   b) SEELONCE MAYDAY
   c) SEELONCE FEENEE

101. When 'designating' a Distress alert by DSC, what options are available?

102. Having sent a DSC alert, when should you transmit a follow-up voice alert?
   a) Immediately
   b) After a few seconds if no response
   c) No need, wait for a response

103. What actions must you take if you transmit a DSC Distress alert by mistake? Give an example.

104. What actions must you take, in accordance with SOLAS, if you receive a Distress alert (by voice or DSC)?

105. When would it be appropriate to send a MAYDAY RELAY message?

106. Give an example of a MAYDAY RELAY message

107. If a Distress message is acknowledged by the coastguard and you are in a position to help, what should you do?
   a) Stand by until asked to render assistance
   b) Call the casualty and offer help
   c) Keep radio silence
   d) Inform the coastguard of your position and ability to help
   e) Ignore the situation – the coastguard will take all necessary actions

Urgency – PAN PAN

108. What is the definition of Urgency?

109. What is the purpose of an Urgency announcement by DSC before a follow-up message is sent by voice?

110. Give an example of a **PAN PAN** voice message.

111. If you get no reply, how often should you repeat the call?
    a) Every minute
    b) Every three minutes
    c) As often as you think necessary

112. Having sent an Urgency announcement, under what circumstances would you upgrade it to a Distress alert?
    a) When you get no reply
    b) To make sure your message is received by another station
    c) If the situation deteriorates to an extent that it fully meets the criteria of Distress

113. Are you *obliged* to always respond to an Urgency announcement?
    a) Yes
    b) No
    c) Only if you may be able to help

114. What message should you send if you need urgent medical advice?
    a) MAYDAY
    b) PAN PAN MEDICO
    c) PAN PAN

Safety – SÉCURITÉ

115. Which of the following would justify a Safety announcement by you?

    a) Your vessel is sinking, but you are in the liferaft

    b) You sight a large baulk of timber in the water which could be a danger to small craft

    c) You are becalmed in a Traffic Separation Scheme

    d) You see a navigational buoy which is out of position

116. Give an example of a Safety announcement.

General

117. Give the meanings of the following:

    a) ATIS

    b) CGOC

    c) CROSS

    d) UTC/UT

    e) NMOC

    f) SOLAS

    g) VTS

    h) SAR

    i) Broadcast

Answers

1. (b) Very High Frequency or VHF (30 – 300MHz).
2. (b) & (c) Shipping and inshore waters forecasts on BBC Radio 4.
3. (b) Also (a), but 'line of sight' is a good rule of thumb.
4. (c) About 15 – 20 miles.
5. (b) About 15 miles.
6. (c) To avoid interfering with other traffic on the same frequency.
7. (c) Global Maritime Distress & Safety System.
8. (a) A1.
9. Within VHF range of at least one coast station using DSC.
10. (a) A1 and A2.
11. No. In such a busy sea area you will almost certainly be able to contact other vessels on VHF to pass on Distress, Urgency or Safety traffic.
12. (f) All of the above.
13. An international network of satellites which relay transmissions from beacons such as EPIRBs and PLBs to SAR authorities.
14. (a) 406MHz.
15. To provide weather and navigational information.
16. (b) English.
17. (b) 200 miles. Some stations may be received up to 400 miles.
18. *Reeds*; Admiralty List of Radio Signals.
19. EPIRB: Emergency Position Indicating Locator Beacon.
    PLB: Personal Locator Beacon.
    SART: Search And Rescue Transponder.
20. Cat 1 EPIRB is mounted on an automatic bracket which releases the beacon as the vessel sinks.
    Cat 2 EPIRB must be removed by hand before it can be activated.
21. UK Beacon Registry.
22. Turn it off and inform the coastguard on Ch 16 as soon as possible giving as much detail as you can.
23. (c) As you get closer to the beacon, the dots extend until they are all-round circles on your screen.
24. (c) Maritime Mobile Service Identity.
25. (a) Ofcom.
26. You only have one chance to get it right! If you confirm the wrong number, the radio will need the attention of a radio engineer.
27. (c) A coast station (eg Coastguard).

28. The radio, and you must apply for a Ship Portable Radio Licence.
29. (c) Ch 70.
30. No! Only for use by DSC.
31. (b) On inland waterways in most countries.
32. No.
33. (a) Collision avoidance – but not all vessels are fitted with AIS.
34. (c) Automatic Transmitter Identification System.
35. '9' is added to the start of your MMSI number. Issued by Ofcom.
36. (b) Turn it up to hear a distinct 'hiss' then back just far enough to suppress it.
37. (c) Press to Transmit. Does exactly what it says.
38. To prevent accidental use. Sometimes you need two hands to press it.
39. (c) Either of the above. On most sets, pressing the Ch 16 button will also select High Power.
40. (c) To maximise range. The higher, the better.
41. (c) Outside UK territorial waters.
42. (b) Ship Radio Licence, issued by Ofcom.
43. (a) & (b) Change of ownership and when equipment is added or removed.
44. (h) All of the above.
45. Yes, with the UK Beacon Registry.
46. (d) Short Range Certificate (SRC).
47. (c) RYA.
48. No, but it is recommended that you make a note of Distress, Urgency and Safety messages which might affect you in the boat's log.
49. (c) 16. The minimum age for taking the exam is 16, but you can attend the course at any age.
50. (c) Must be *directly* supervised. (Common sense dictates that in an emergency anyone on board should use the radio to call for help if the SRC holder is incapacitated.)
51. (b), (c) & (d) No unofficial 'callsigns'!
52. (b) No.
53. (c) When stood down by the coordinating authority.
54. (c) Use 'All stations'. Broadcasting 'blind' is not allowed.
55. (c) Listen to ensure the channel is not busy.
56. (d) You may block other transmissions. (b) & (c) are also relevant.
57. (d) & (e) Radio checks are rarely necessary.

58. *First* on your list should be (d), DSC; *last* on your list should be (a), the coastguard on Ch 16. The other options are all acceptable, but testing with a handheld set may not prove much because of limited range.
59. (b) I have finished my transmission and am waiting for your reply.
60. (c) I do not expect a reply from you.
61. (c) *Never* say 'Over and out'!
62. (b) Received. However, 'Roger' is still often heard.
63. (c) Only authorised stations: marinas, coastguard, etc.
64. (e) Say again all after/before ...
65. (d) Station calling [my vessel], say again your name. Over.
66. (c) Two minutes. Try three times before waiting longer.
67. Ch 6, 8, 72 & 77
68. 9, 10, 15, 17, 69 & 73. Note that Ch 10 is used for some MSI broadcasts. Ch 15 and Ch 17 are both limited to low power.
69. A channel that uses two frequencies, one to transmit and the other to receive.
70. A channel that uses the same frequency to transmit and to receive.
71. (d) Ch 80 for most UK marinas.
72. (c) Distress, Urgency and Safety working. Also used as the calling channel (a), but keep transmissions as short as possible.
73. (c) Ch 13. If no response, try Ch 16 to establish contact then change to Ch 13 or other working channel.
74. Usually Ch 10, 63, 63 & 64. Prior announcement on Ch 16.
75. (b) *Reeds*. (c) & (d) also possible.
76. Because Ch 80 is a duplex channel, and most yachts and small craft have only one VHF aerial (semi-duplex radios).
77. Ch 70 is dedicated to DSC transmissions only.
78. Ch 0 is used by the coastguard and emergency services.
79. Ch M is used by some marinas and yacht clubs. Not to be used outside UK territorial waters.
80. No – see answer 79 above.
81. Ch 16 (to monitor Distress and calling); Ch 13 (to monitor bridge-to-bridge traffic). Possibly the local VTS channel to monitor shipping movements and other information. In the southern North Sea this might be, for example, the Sunk VTS on Ch 14.
82. (a) & (b) Listen to Ch 16 to make it is not busy. Think what you are going to say

83. (d) On the initial call you might say the name up to three times. Once contact is made, once is ample. To avoid any confusion, remember to say your name once before each transmission ('This is ...').

84. (b) The other vessel. If she does not select a working channel, suggest one yourself having checked it is not busy.

85. Vessel's name; callsign; MMSI number; where from (and departure time); where to (and ETA); number of persons on board. Remember to specify the time zone if appropriate.

86. (c) (possibly) Speak normally, but particularly distinctly and slightly slower than usual. If you have a naturally low-pitched voice, raising it slightly might help.

87. (c) Slightly to one side and about 2 – 5 cm away.

88. Keep it short!

89. (d) Individual.

90. (b) MMSI number.

91. (c) Never use Ch 70 for voice communications.

92. (b) & (c) Once you are practised in using DSC, it can be quicker and saves any negotiation about a working channel.

93. (a) Charlie Oscar Whiskey Echo Sierra
Look up the others, if necessary!

94. A *vessel* or *person* [must be in] *grave and imminent danger.*

95. (a) Any danger you may encounter is not 'imminent'
(c) A dog is not a person – whatever you might think!
(f) You may have a long and tedious wait, but you are not in 'grave danger'.
Note: (d) might also apply. A MOB will usually justify a MAYDAY call, but in calm conditions, warm water and with an uninjured casualty who you are confident of rescuing quickly, you might consider there to be no 'grave and imminent' danger.

96. (c) Ch 16.

97. Identity (name, MMSI number); position; nature of distress; assistance required. For full format, see Chapter 9 and next answer.

98. 'MAYDAY MAYDAY MAYDAY
This is yacht *Voyager*, *Voyager*, *Voyager*
MMSI number 235 123 456
Callsign: Mike Whisky November Five
MAYDAY *Voyager*
My position is [see answer 99 on the following page]
My vessel is taking in water and sinking

There are two adults and one child onboard
I require immediate assistance
Over'

99. Latitude and Longitude
Range and bearing *from* a charted object
General description (eg 'About 10 miles east of Spurn Head')

100. (a) MAYDAY RELAY – used to pass on the details of a Distress situation to other stations. Normally to the coastguard, but may be to 'All ships' if you get no response.
(b) SEELONCE MAYDAY – used to impose radio silence (a specific channel may be nominated) during Distress working.
(c) SEELONCE FEENEE – used to lift radio silence.

101. Undesignated, Fire, Flood, Collision, Grounding, Capsizing, Sinking, Adrift, Abandoning, Piracy and Man overboard.

102. (b) Wait a few seconds to allow another station to respond.

103. Cancel the alert immediately. Read your radio's manual to see if there are any particular procedures (eg turn it off then back on), then send a message to 'All stations' on Ch 16 giving your vessel's name, callsign and MMSI number followed by: 'Cancel my Distress alert. I say again, cancel my Distress alert'.

104. In theory, you must respond and offer any assistance you can. However, in a small craft such as a yacht, unless you are in a position to help, wait a few minutes to see if a more appropriate station (eg the coastguard) responds. If so, maintain silence on any channels being used. If no response, relay the Distress alert to 'All stations' by voice on Ch 16.

105. If you know or suspect that another vessel is in Distress and/or there has been no response to a MAYDAY call.

106. 'MAYDAY RELAY, MAYDAY RELAY, MAYDAY RELAY
Dover Coastguard, Dover Coastguard, Dover Coastguard
This is yacht *Voyager*, *Voyager*, *Voyager*
MMSI: 235 123 456
Received following MAYDAY from yacht *Seabird*
[Repeat Distress call or describe the situation]
Over'

107. (d) Inform the coastguard of your position and ability to help. Include your ETA at the casualty if appropriate.

108. When you have a *very urgent message* concerning the *safety* of a *ship or person*.

109. A DSC Urgency announcement will alert all receiving stations by sounding an alarm. It does *not* include your position, and a follow-up PAN PAN call must be made on Ch 16 to 'All stations'.

110. 'PAN PAN, PAN PAN, PAN PAN
This is yacht *Voyager*, *Voyager*, *Voyager*
MMSI number 235 123 456
My position is [see answer 99 above]
I have been disabled and am drifting towards the Shingles Bank
I require a tow into the Solent
Two adults on board
Over'

111. (c) As often as you think is necessary, but allow time for a response.

112. (c) If the situation deteriorates so that it fully meets the criteria of Distress.

113. (c) You should only respond if you are able to offer assistance.

114. (c) PAN PAN. The old prowords of PAN PAN MEDICO are no longer used.

115. (b), (c) & (d)   If you have taken to the liferaft, an Urgency announcement or Distress alert would be more appropriate, depending on the circumstances.

116. 'SÉCURITÉ, SÉCURITÉ, SÉCURITÉ
All stations, all stations, all stations
This is yacht *Voyager*, *Voyager*, *Voyager*
MMSI number 234 123 456
My position is [see answer 99 above]
Semi-submerged shipping container see in position [see answer 99 above]
Over'

117. (a) Automatic Transmitter Identification System.
(b) Coastguard Operations Centre (formerly known as an MRCC).
(c) Centre Régional Opérationnel de Surveillance et Sauvetage (French equivalent of a UK CGOC).
(d) Coordinated Universal Time/Universal Time (equivalent to GMT).
(e) National Maritime Operations Centre
(f) Safety of Life at Sea [Convention]
(g) Vessel Traffic Service
(h) Search and Rescue
(i) Sending a message without specifying who it is intended for (illegal!).

Appendices

Appendix A

Overheard on VHF

Here are just a few examples – all true – of VHF transmissions which are either amusing, include unfortunate wording or simply highlight poor VHF procedure. I have not included any of the myriad transmissions using *'Over and out'* (or other ubiquitous howlers), nor have I even attempted to reproduce interminable chats between fishermen who understandably, if illegally, use VHF to while away the long night watches.

1. Container ship about to leave harbour passing details on Ch 12 to Southampton VTS:
   VTS: **'… and do you have any hazardous cargo on board, Sir?'**
   Ship's master: **'Only my wife.'**

2. Conversation between a yacht and the coastguard:
   CG: **'What is your position?'**
   Yacht: **'I am on the floor by the chart table and unable to move.'**

3. A yacht trying to call Port Control:
   Yacht: **'Port Control, Port Control this is yacht *Calamity**, *Calamity*, over.'**
   [* Name changed to save blushes.]
   Port Control: **'*Calamity* this is Port Control. Over.'**
   Yacht: **'Port Control, Port Control this is yacht *Calamity*, *Calamity*. Over.'**

Port Control: **'*Calamity*, I say again, this is Port Control. Over.'**

Yacht: **'Port Control, Port Control this is yacht *Calamity*, *Calamity*. Over.'**

Port Control: **'*Calamity*, I say again, this is Port Control. Over.'**

Yacht: **'Port Control, this is *Calamity*, I have put my hearing aids in now. Over.'**

4. Conversation between ship and a doctor ashore:

Doctor: **'Is the casualty breathing? Over.'**

Ship: **'No. He no breathe; he no move. Over.'**

Doctor: **'Then be advised the casualty is dead. Over.'**

5. Call on Ch 16 between two yachts:

Yacht A: **'Are you going into Lymington?'**

Yacht B: 'Yes, we should be alongside in about 30 minutes.'

Yacht A: **'Shall we meet at that new place on the High Street?'**

Yacht B: 'Good idea.'

Solent coastguard: **'Ladies, please discuss your luncheon arrangements on a working channel. Out.'**

6. Yacht trying to contact another:

'Hello, hello. Are you there? Come in, please. Hello?'

Appendix B

Check-off list for Distress alerts

DISTRESS PROCEDURE (DSC)

 This list must be adapted for your particular radio.

IF YOU ARE IN **IMMEDIATE** DANGER:

1. **LIFT UP** spring-loaded **Distress** cover on the radio

2. **PRESS AND HOLD** red 'Distress' button until message is sent (ie when display stops flashing – about five seconds)

3. **WAIT** about 15 seconds for an acknowledgment, then ...

4. SEND **MAYDAY** call by voice

   Note: The radio will automatically resend the **Distress** call every four minutes until it is acknowledged

   IF YOU HAVE TIME:

5. **PRESS** [CALL/Menu] button

6. **ROTATE** [CH] knob to select '**Distress Alert Msg**'

7. **PRESS** button under [Select]

8. **PRESS** button under [Nature]

9. **ROTATE** [CH] knob to select the type of distress:

   Fire
   Flood
   Collision
   Grounding
   Capsizing
   Sinking
   Adrift
   Abandoning
   Piracy
   MOB

10. **PRESS** button under [Select]

11. **SEND** DSC message – **Steps 1 to 4 above**

Appendix C

Pro forma for **MAYDAY** calls

Additional data to be inserted in the red box opposite:

- Your vessel's name (three times)
- Your MMSI
- Your vessel's name after MAYDAY
- Your position
- The nature of your Distress
- The number of people on board

This pro forma must be adapted for your particular vessel

1. **FIND THE POSITION** – and *write it down*

Latitude and Longitude (from the radio, chart plotter or chart)

   Or *Bearing and Range* (from a charted object)

   Or *General area* ('About ten miles south of the Needles')

2. **PRESS** red [16] button on the radio

3. **CHECK** that '25W' is showing on top line of display. If not, press [H/L] button.

4. **PRESS** transmit button on microphone; *speak slowly and clearly:*

MAYDAY, MAYDAY, MAYDAY

This is:

MMSI number is:

Callsign is:

MAYDAY

My Position is:

(State nature of distress and number of people onboard)

Over

---

5. **RELEASE** transmit button; listen for a response. If none, repeat MAYDAY call as necessary.

Appendix D

MMSI NUMBERS

Station Name	MMSI	Callsign	Notes

VHF CHANNELS

Station	Working Channel(s)	Notes
Distress & Urgency	16	
Small Craft Safety (UK)	67	
Bridge-to-Bridge	13	For navigational safety
Ship-to-Ship (primary)	6, 8, 72, 77	aka Inter-ship
Ship-to-Ship (alternative)	9, 10, 15, 17, 69, 73	Ch 15 & 17 are low power
NCI	65	
MSI Broadcasts	10, 62, 63, 64	
Marinas	80	Most UK marinas

Index

Titles of Interest

Adlard Coles Book of Electronic Navigation

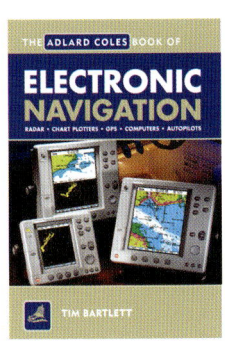

Melanie Bartlett
ISBN 9780713657159

Using plain English to strip away the jargon, this book sets out to demystify the technology behind modern marine electronics, show how the various systems work and explain how to get the best out of the equipment available. It is aimed at anyone with an interest in navigating small craft, but particularly those working towards their Day Skipper, Yachtmaster and other RYA courses.

First Aid At Sea 7th edition

Douglas Justins and Colin Berry
ISBN 9781472953414

Providing an easy-to-navigate instant guide to emergency first aid for all seafarers, this book includes a colour-coded thumb index of emergencies for quick reference, a concise description of medical conditions with a prioritised list of treatments, ringbinding and waterproofed pages to withstand use at sea, and is fully updated in line with guidelines and best practice.

GMDSS: A User's Handbook 6th edition

Denise Bréhaut
ISBN 9781472945686

Anyone with GMDSS equipment on board their vessel will need an operator's licence. This book has long been an invaluable reference for exam candidates and all marine users, from novice to advanced. It explains the operation of the system as a whole and the procedures involved, as well as covering the syllabi of the General Operator's Certificate (GOC), the Restricted Operator's Certificate (ROC), the Long Range Certificate (LRC) and the Short Range Certificate (SRC). The 6th edition of GMDSS incorporates all the changes to the regulations that came into force in 2009 as well as the 2016 system updates.

Reeds Skipper's Handbook 7th edition

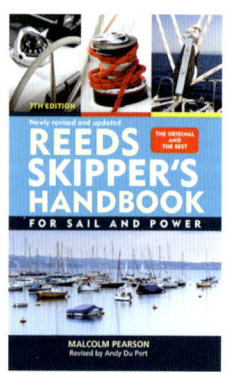

Malcolm Pearson (Fully revised by Andy Du Port)
ISBN 9781472972163

The bestselling aide-memoire of everything a boater needs to know at sea. A handy pocket size, it is packed with a wide range of information in a concise form and is a must for anyone going to sea in any size of boat – be they novice or old hand.

Tens of thousands of skippers and crew have found it invaluable as a memory jogger and refresher whether at sea or on land and it is frequently recommended by Yachtmaster Instructors as a quick reference guide and as a revision aid for anyone taking their Day Skipper or Yachtmaster certificates.

Skipper's Onboard Emergency Guide

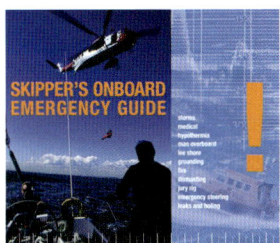

Hans Donat
ISBN 9780713684957

Providing practical, accessible advice on the most common emergencies requiring rapid action, this handy guide covers storm tactics, man overboard, gear failure, fire, dismasting, sinking, first aid and much more. Filled with checklists, sequential action points and helpful diagrams, this is an invaluable cockpit reference for anyone finding themselves in an emergency at sea.

The Yacht Owner's Manual

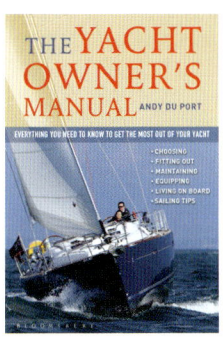

Andy Du Port
ISBN 9781472905482

Owning a boat involves sailors developing a whole new skill set and taking on a raft of new responsibilities, but this essential handbook takes the stress out of what should be an enjoyable, rewarding next step, and shares a wealth of practical advice on what might initially seem daunting new challenges.

Filled with colour photos and diagrams throughout, this comprehensive guide is indispensable for new skippers who have learned to sail at a sailing school or on a friend's boat and are ready to take the next step.

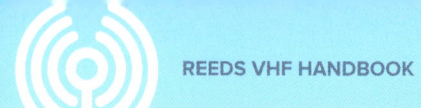

Your First Channel Crossing

Andy Du Port
ISBN 9781408159088

After day sailing, the next step for the adventurous yachtsman or motorboater is their first open sea crossing, and for many that invariably entails crossing the English Channel. This invaluable guide walks the first timer through their first Channel crossing step by step, tackling all the aspects that have to be taken into account before embarking, as well as managing on the trip, in the order in which they will be encountered.

Encouraging, engaging and practical, by covering each stage in precisely the amount of detail required for the first-timer, it is like having an instructor or mentor on board.

Photo credits

Getty Images pp.39; 54; 62; 85; 94; 99; 146

Shutterstock pp.2; 6–7; 12–13; 18–19; 32–3; 42–3; 48–9; 56–7; 59; 64–5; 68–9; 76–7; 80; 81; 87; 88–9; 92–3; 96–7; 100–1; 104–5; 110–11; 113; 114–15; 142–3